# Introduction to Digital Systems

# Introduction to Digital Systems

John Crisp

OXFORD BOSTON JOHANNESBURG MELBOURNE NEW DELHI SINGAPORE

Newnes
An Imprint of Butterworth-Heinemann
Linacre House, Jordan Hill, Oxford OX2 8DP
225 Wildwood Avenue, Woburn, MA 01801-2041
A division of Reed Educational and Professional Publishing Ltd

A member of the Reed Elsevier plc group

First published 2000



**British Library Cataloguing in Publication Data**
A catalogue record for this book is available from the British Library

ISBN 0 7506 4583 0

**Library of Congress Cataloguing in Publication Data**
A catalogue record for this book is available from the Library of Congress

Composition by Genesis Typesetting, Rochester, Kent
Printed and bound in Great Britain by Biddles Ltd, www.biddles.co.uk

# Contents

# Preface

Modern life depends on a few basic requirements and of these, one of the most important is digital electronics.

In fact, it is difficult to imagine what would still be available to us if digital systems were removed. We would lose our computers, telephones, televisions, radios, CDs and microwave ovens. Our transport system could not include modern ships, aircraft, trains and even many cars.

As the days pass, we become more dependent on digital systems. On our planet digital devices outnumber us by more than a billion to one! It's a good job they are friendly.

The purpose of this book is to give a worry-free introduction to the world of digital electronics. It starts at the beginning and does not assume any previous knowledge of the subject, and new topics are fully explained as they are introduced.

**John Crisp**

## Digital and analogue

There are two types of light control found in our homes: light switches and light dimmers. With a light switch, the light is either on or it is off. The light switch is called a 'digital' control since it changes state completely with no intermediate values. On the other hand, a light dimmer will allow all light values between on and off. This type of control is called an 'analogue' control.

This distinction between analogue and digital can be applied to most devices and processes. For example, a mercury thermometer is an analogue device since the mercury can creep up the tube reaching all values in turn.

Time is analogue and, as it creeps by, we can use a calendar or a clock to keep track of it. A calendar is a digital device. The calendar indicates the date of 2 October 1999 for the whole of the day. Twelve hours into the day we don't say the date is 2½ October 1999. No, we wait for 24 hours then leap forward to 3 October 1999. In a similar way, a digital clock shows the same time for a while then jumps forward to the next value. An analogue clock however – that is, one with moving hands – shows time passing as a smooth movement of the hands.

The digital way of indicating the passage of time is shown in Figure 1.2.

**Figure 1.2**
One way of measuring time

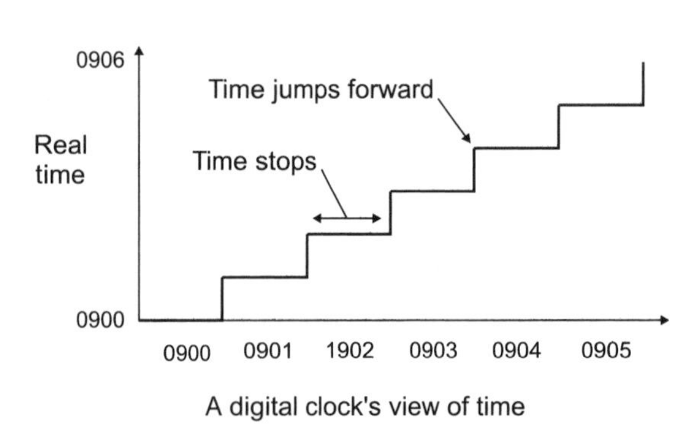


## What is a system?

Let's start by looking at one.

The word 'system' is used to describe any organization or device that includes three features.

A system must have at least one input, at least one output and must do something, i.e. it must contain a process. Often there are many inputs and outputs. Some of the outputs are required and some are waste products. To a greater or lesser extent, all processes generate some waste heat. Figure 1.3 shows these requirements.

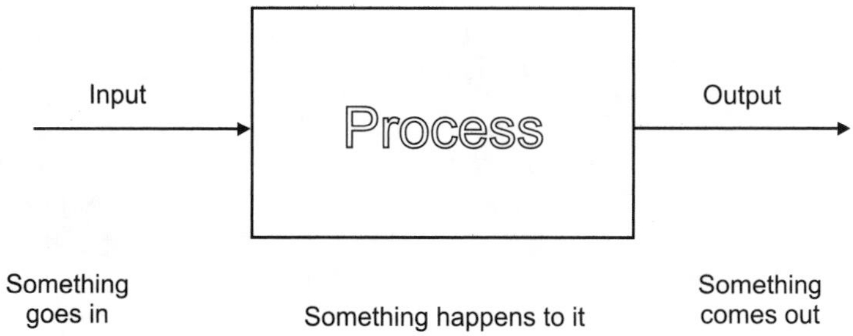

**Figure 1.3**
The essential requirements of a system

Input → Process → Output

Something goes in | Something happens to it | Something comes out

A wide range of different devices meets these simple requirements. For example, machinery does something, or performs a function. It also requires inputs like fuel to make it work, and it always has outputs. They may be wanted outputs like movement, or they may be waste products like noise and pollution. Figure 1.4 shows the main parts of a monster-truck system.

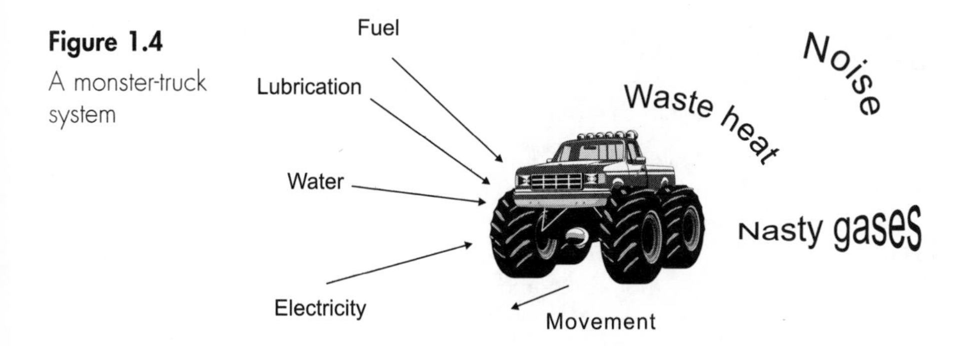


**Figure 1.4**
A monster-truck system

The monster truck contains other systems within it. In Figure 1.4, we add electricity as a required input to start the engine and provide the lights and the instruments, but thereafter the battery is recharged by the engine. There must, then, be an electrical system at work – so it is quite possible for systems to contain smaller systems. In a similar way, a monster truck is just a part of the transport system.

## Digital systems

Returning to our light switch, the switch is described either as being ON or OFF.

In digital electronics, we normally use the digits 1 and 0 to represent these two states. It is our choice whether to use the digit 1 to represent ON or to represent OFF. The most usual convention is to use 1 to represent ON and the 0 to represent OFF.

In fact, the use of 1 and 0 can be employed to represent any two states which are opposite in the chosen sense. For example, given the choices black/white, true/false, up/down, left/right, we could decide to define black as 1, which would make white = 0. Remember that it was our choice, so we could have decided to make black = 0 and white = 1.

If there is already a convention in use it is easier for everyone if we stay with it, but in cases where there is no convention, or we don't know it, we are free to make our own choice. In this case, however, we must be very careful to state our definition so that other people can understand our intentions.

In these days of global markets, the on/off switch is often marked with 1 and 0 to avoid the necessity of printing the front panels with 'on' and 'off' in many different languages. Millions of people are therefore using digital symbols without realizing it.

## Binary systems

A digital system that has only two states is called a 'binary' system.

If we stood on the side of a road and watched the traffic go by, we could divide the vehicles into two categories. We may decide on 'cars' and 'lorries', and settle down with a note pad to record the first six vehicles. Around the corner come a car, a combine harvester, a car, a steamroller, a motorbike and a tractor.

The choice of categories can clearly be improved. To record the traffic stream in a binary from, we need to write down a definition in such a way that only two possibilities exist and all types of vehicle must fit into the scheme. A possible solution is to divide all vehicles into the two groups that we could call 'cars' and 'not cars'.

When we write down binary results it is often easier to use the numbers 0 and 1 to represent the two possibilities, so we would start by saying what the number 1 means and then we can assume that 0 means the opposite. How we define things is up to us, but once decided, we must stick to it.

Let's assume that a car is given the value 1 and therefore 0 is used for 'not cars'.

Our record book would show the first six vehicles as:

| Vehicle | 1st | 2nd | 3rd | 4th | 5th | 6th |
|---|---|---|---|---|---|---|
| Vehicle type | 1 | 0 | 1 | 0 | 0 | 0 |

So we can see that we have a car which is given the binary value 1, followed by a 'not car' that is recorded as a 0, then another 1, followed by three 'not cars', 0, 0 and 0.

If we now ask the question 'How many vehicles were cars?', the answer is easily found by counting the number of 1s that have been listed.

## Logic states

We often use digital electronics for control or decision-making circuits, and for this reason the circuits are often called 'logic' circuits and the 1 and 0 conditions 'logic states'.

If we wished to design a digital circuit that could count the number of cars that pass in a whole day we could (possibly) design a circuit that would detect whether a vehicle was a car or not, and hence give a series of logic 1s and 0s at the output.

The next step would be to design a circuit that could count the number of 1 states that occur at the output.

Digital circuits that are designed to count are extremely fast and reliable, but do not usually use our familiar method of counting.

We will look as the alternative systems in the next chapter.

### Quiz time 1

In each case, choose the best option.

**1 The monster-truck system in Figure 1.3 shows**:

(a) only three outputs.
(b) only one output.
(c) four outputs.
(d) seven inputs.

**2 All systems must include:**

(a) an input, an output and a process.
(b) something to do with a form of transport.
(c) electricity.
(d) fuel, water and electricity.

### 3 All systems generate:

(a) movement.
(b) light.
(c) waste heat.
(d) waste gases.

### 4 The brightness level of daylight:

(a) varies in an analogue way.
(b) must be represented as a 0 state at midnight.
(c) varies in a digital way.
(d) changes its logic state at noon each day.

### 5 When describing a digital switch, the 1 state:

(a) must refer to its ON state.
(b) must refer to its OFF state.
(c) can be defined as either state but usually refers to its OFF state.
(d) can be defined as either state but usually refers to its ON state.

# 2

# It's numbers that count

This chapter deals with one of the ways in which a digital circuit can handle numbers.

Counting is not an easy thing to learn. If you think back, it took the combined efforts of parents and teachers over the first 6 or 7 years of our lives before we could feel confident with counting. We were only learning one way of doing it. Now, in this one chapter, we are going to look at three or four ways – so don't worry if you need to read it twice.

Get the idea of it and refer back later for the details if you feel the need.

First we will zoom back to primary school to see how we count in the 'normal' system. We all started counting using our fingers, and people have been doing this for so long that the word 'digit' means both finger and number.

## Denary, or counting with ten fingers

This is 'normal' counting. We start off with 0 then count up as shown.

0
1
2
3

4
5
6
7
8
9

At this point, we run out of symbols. Notice that there are ten different symbols – hence the name denary.

To continue the count, we put a 1 in the next column to the left to show that we have used all the symbols once, and reset the right-hand column to 0. Then we go again:

10
11
12
13
etc.

When we have 19, we reset the right-hand digit to 0 and increase the left-hand column to 2. This gives us the number 20.

We continue to count in this manner until we reach 99, whereupon we reset both columns to 00 and introduce a 1 in the next column to the left to give us 100.

We have spelled out denary counting in some detail to set the scene for the other systems.

## Some words

To make our normal counting seem more difficult than it is, we can introduce some technical words – there is nothing quite as good as a few technical-sounding words to impress people with our expertise.

Because we use ten different digits, we call this system 'denary'. We also call it the 'decimal' system. The words 'base', 'radix' or 'modulo' indicate the number of different digits used.

So our everyday system could be referred to in any of the following ways:

counting in denary
counting in decimal
counting using base-10
counting using a radix of 10
counting using modulo-10, sometimes abbreviated to mod-10.

Impressive, isn't it?

Since our logic circuits can only give the output result as a 0 or 1, we must design them to count using only these two digits. This system is called 'binary' counting.

## Binary, or counting with two fingers

Binary uses only two digits, 0 and 1.

'Binary digit' is referred to so often that we have abbreviated the term to 'bit' (derived from BInary digiT).

## Counting in binary

We just use the same rules as we applied in denary counting, except that we must remember to use only the two digits.

We start to count:

0
1

Now we have run out of digits, so we reset the right-hand column to 0 and add a 1 in a new column to the left just as we did with denary.

We then go:

10
11

and reset these columns to zero and add a new column to the left to give:

100
101
110
111

and then

1000
1001
1010
etc.

Let's compare binary with denary.

| Binary | Denary |
|---|---|
| 0 | 0 |
| 1 | 1 |
| 10 | 2 |
| 11 | 3 |
| 100 | 4 |

| Binary | Denary |
|---|---|
| 101 | 5 |
| 110 | 6 |
| 111 | 7 |
| 1000 | 8 |
| 1001 | 9 |
| 1010 | 10 |

Tip: If we read a binary number like 100 as 'one zero zero' rather than 'a hundred', it will help us to avoid mistakes.

Notice that in the denary system, the columns will have values of units, tens, hundreds etc. as we move from right to left.

Mathematically, these values are written as $10^0$, $10^1$, $10^2$ etc.

Similarly, with the binary count the columns have values of units, twos, fours, eights, sixteens etc. as we go from right to left across the columns.

These are written as $2^0$, $2^1$, $2^2$, $2^3$ etc.

The denary number 239 means the total of 9 units, plus 3 tens, plus 2 hundreds. In a similar way, 1011 in binary means 1 unit, plus 1 two, plus 0 fours, plus 1 eight. This gives a total that we would call eleven in denary.

We have to be careful here. Have a look at Figure 2.1.

**Figure 2.1**
Numbers can be confusing

What number is this?

If we write 10 on the page, what do we mean? It could be ten if we assume it to be a denary number, or it could equally well be 10 in binary, which is two in denary.

The way out of this problem is to write a small number called a subscript after the last digit to show the base of the number system. If it were a binary number we would add the subscript 2 so the number would be written as $10_2$, but if it were a denary number we would write $10_{10}$.

## When are subscripts used?

A maths purist may well insist on the base being shown on all occasions, but in the real world this is not done. We never quote our age on an application form as $25_{10}$. No, the common-sense approach is to use subscripts whenever there is a likelihood of our intentions being misunderstood.

Since digital circuits operate in binary and we work in denary it is important for us to be able to convert numbers between the two systems.

## Changing binary numbers to denary

### Sorting out the columns

When we write numbers we always put the lowest number value in the right-hand column and this is called the 'least significant bit', abbreviated to LSB. The highest value number is in the left-hand column and is the 'most significant bit', the MSB.

A denary number like 2357 can be written in columns labelled:

| $10^3$ | $10^2$ | $10^1$ | $10^0$ |
|---|---|---|---|
| 2 | 3 | 5 | 7 |

Each column is ten times larger than the one to its right.

When we come to binary we must remember that the column values increase in powers of 2, so the columns are twice as large as the one to the right.

| $2^7$ | $2^6$ | $2^5$ | $2^4$ | $2^3$ | $2^2$ | $2^1$ | $2^0$, or |
|---|---|---|---|---|---|---|---|
| 128 | 64 | 32 | 16 | 8 | 4 | 2 | 1 |

### Converting by columns

To change a binary number like 1011001 into a denary number, we can just write down the column values and enter the binary number.

| 64 | 32 | 16 | 8 | 4 | 2 | 1 |
|---|---|---|---|---|---|---|
| 1 | 0 | 1 | 1 | 0 | 0 | 1 |

This means that we have:

$1 \times 64 = 64$
$0 \times 32 = 0$
$1 \times 16 = 16$
$1 \times 8 = 8$
$0 \times 4 = 0$
$0 \times 2 = 0$
$1 \times 1 = 1$

Once we see that $0 \times$ anything is always zero, we can simplify the process considerably by just listing the columns where the binary value is one. The above list can be simplified to $64+16+8+1=89$, so the final result is $1011001_2=89_{10}$.

**Remember the subscripts**

Since we are always using both binary and denary in these conversions, be very careful to avoid misleading answers. For example, having converted 10 in binary to 2 in denary, don't write it as $10=2$.

**Example**

Convert $11010110_2$ to denary.

Write out the column values:

| 128 | 64 | 32 | 16 | 8 | 4 | 2 | 1 |
|---|---|---|---|---|---|---|---|

Write out the binary number in the columns:

| 128 | 64 | 32 | 16 | 8 | 4 | 2 | 1 |
|---|---|---|---|---|---|---|---|
| 1 | 1 | 0 | 1 | 0 | 1 | 1 | 0 |

List all the column values that hold a binary value of 1, then add them:

$128+64+16+4+2=214$

Answer: $11010110_2=214_{10}$.

## Converting denary numbers to binary

If we needed to know what the number five was in binary, we could just count up from zero in binary until we got to the required number. It would be a long and tedious process to use this method to find the binary equivalent of 500 – and we would probably get it wrong.

**Calculator note**: Many scientific calculators can do the conversion of denary to binary for us. To convert 52 into binary, we would simply enter the number 52 then switch the calculator to binary mode and the result would be shown. Unfortunately, they are limited to quite low numbers by the number of digits able to be seen on the screen.

Here is a better way. Let's assume that we want to convert $25_{10}$ to binary.

Step 1: Write down the number.

25

Step 2: Divide it by two and write down the whole number part of the answer; next to it write the remainder, which will always be one or zero.

| | |
|---|---|
| 25 | |
| 12 | 1 |

Step 3: Now do the same thing again. Divide the 12 by two and write down the answer 6, and the remainder (0) next to it.

| | |
|---|---|
| 25 | |
| 12 | 1 |
| 6 | 0 |

Step 4: Do the same again; divide the six by two, which gives 3 and a remainder of 0.

| | |
|---|---|
| 25 | |
| 12 | 1 |
| 6 | 0 |
| 3 | 0 |

Step 5: Divide the three by two, which gives 1 with a remainder of 1.

| | |
|---|---|
| 25 | |
| 12 | 1 |
| 6 | 0 |
| 3 | 0 |
| 1 | 1 |

Step 6: Be careful with this last step. When we try to divide the last one by two, the answer is 0 with a remainder of 1.

| | |
|---|---|
| 25 | |
| 12 | 1 |
| 6 | 0 |
| 3 | 0 |
| 1 | 1 |
| 0 | 1 |

Step 7: That is the end of the dividing now that we have got to zero.

The binary number now appears in the remainder column. To get the answer, read the remainder column from the bottom UPWARDS, as in Figure 2.2.

**Figure 2.2**
Reading the binary number

| 25 | Remainders |
|---|---|
| 12 | 1 |
| 6 | 0 |
| 3 | 0 |
| 1 | 1 |
| 0 | 1 |

**11001**

Read the remainders from the bottom upwards so $25_{10} = 11001_2$

## Summary

1 Divide the denary number by two. Write the whole number result underneath and the remainder in a column to the right.
2 Repeat the process until the number is reduced to zero.
3 The binary number is found by reading the remainder column from the bottom upwards.

Here is one for you to try. If you get stuck, the solution is given below.

**Example**

Convert $86_{10}$ to a binary number.

**Answer**

| 86 | |
|---|---|
| 43 | 0 |
| 21 | 1 |
| 10 | 1 |
| 5 | 0 |
| 2 | 1 |
| 1 | 0 |
| 0 | 1 |

So $86_{10} = 1010110_2$.

**Note**: If the denary number is even, the binary number always ends with a zero.

For additional practise, just choose any denary number, convert it to binary and then convert your answer back to denary. With some luck you should get back to the same number.

## Quiz time 2

In each case, choose the best option.

**1 The binary number 11100110 is equivalent to the denary number:**

(a) 103
(b) 255
(c) 230
(d) 102

**2 The number $645_{10}$ is equivalent to the binary number:**

(a) 1001100011
(b) 1010000101
(c) 1011101001
(d) 101 x 0001101

**3 Denary:**

(a) has a base of 9.
(b) is called modulo counting.
(c) can only count up to 9.
(d) has a radix of 10.

**4 A binary number:**

(a) must include the numbers 1 and 0.
(b) always has more digits than the equivalent denary number.
(c) 100 is called a 'hundred'.
(d) with 5 bits cannot represent a number greater than $31_{10}$.

**5 A bit is:**

(a) a small piece of cake.
(b) a single binary digit.
(c) derived from the words 'Binary In Transit'.
(d) used as to describe a single digit in binary or in denary.

# 3

# Binary arithmetic

## Adding binary numbers

This follows the same process as we normally use for addition in the denary system. Remember that we are working in binary so if we add, for example, 1 + 1, the answer in binary is $10_2$ not 2. Keep saying 'binary' to yourself. It is so very easy to slip back to denary without realizing it.

We will now look at an example in detail and show the step by step sequence.

## Adding two binary numbers

### Example

We are going to add two binary numbers.

```
1011
 111 +
----
```

Step 1: Starting with the right-hand column, we add the 1 and the 1 to give 10. (Binary, remember!). The 0 goes in the right-hand column as the answer and the 1 goes under the next column as a 'carry'.

It now looks like this:

```
1011
 111 +
----
   0
----
  1   carry
```

Step 2: We can now add the next column. This is 1 + 1 + the carry. In binary, this gives the value 11, or answer 1 and carry 1.

```
1011
 111 +
----
  10
----
 11   carry
```

Step 3: In the third column we have 0 + 1 + carry, which in binary results in 10, so 0 goes in the answer and 1 goes in the carry.

```
1011
 111 +
----
 010
----
111   carry
```

Step 4: In the last column we only have a single 1 plus the carry, giving a result of 10 in binary. The result is therefore a zero answer and a 1 in the carry.

```
 1011
  111 +
 ----
 0010
 ----
 1111   carry
```

Step 5: The final carry produces another column on the left-hand side to give us a final answer of 10010.

So 1011 + 111 = 10010 in binary, or $1011_2 + 111_2 = 10010_2$.

**Example**

Add $1001_2 + 101_2$.

Your answer should be $1110_2$. If you don't agree, the working is shown below.

Step 1: Adding the right-hand column gives 1 + 1 = 10, which is an answer of 0 and 1 in the carry.

```
1001
 101 +
----
   0
----
  1   carry
```

Step 2: Adding the second column gives 0 + 0 + 1 in the carry, which gives an answer of 1 and no carry.

```
1001
 101 +
----
  10
----
  1   carry
```

Step 3: The third column gives 0 + 1 and so we have an answer of 1 and no carry.

```
1001
 101 +
----
 110
----
  1   carry
```

Step 4: The last column only contains a 1 and since there isn't a carry, the result is just 1, giving the answer of 1110.

```
1001
 101 +
----
1110
----
  1   carry
```

## Adding more than two binary numbers

With three or more binary numbers to add, you may find it easier to add the first two numbers and then add the third, then the fourth, and so on. This is the way that a digital adding circuit would tackle the problem. You could, of course, simply list all the numbers and add them up at once to give a final total, but you will need a little more practice since the 'carry' situation tends to become more confusing. Fortunately, we seldom have to add more than two binary numbers.

### How many bits should we use?

Have a look at these binary additions:

1100 + 1001 = 10101
1100 + 10 = 1110

In the first example, the two numbers being added each have 4 bits but the answer has 5 bits. In the other case, the answer is a 4-bit number but one of the numbers being added has only 2 bits.

This variation in the number of bits is easy to handle when we are simply adding numbers by hand, but if we design an electronic circuit to do the job it is not so easy.

Each bit must be entered into the circuit by applying a voltage to an input wire, and the result appears on another set of wires. Now, we can't have wires disappearing when they are not needed and suddenly appearing when an extra bit is needed.

So digital circuits must be designed to work with a set number of digits (bits), and it is the designer's responsibility to decide how many bits and therefore how many wires to use.

The criteria are:

1 The more bits used, the more expensive is the design.
2 Increasing the number of bits in a binary number increases the maximum number that can be represented. Every extra bit will double the maximum number.
3 It may have to fit in with other devices like microprocessors, which use numbers with a fixed number of bits – generally 4, 8, 16, 32, 64 and 128 bits at the present time.

In some of our earlier examples, we had to create an extra column on the left-hand side due to the final carry. This must be catered for when designing the digital circuitry.

Have a look at this addition:

```
  11
  10 +
 101
```

If we decided to operate with only 2 bits, the left-hand column in the answer would be ignored since only two wires would come out of the adding circuit. The sum would then look like this:

```
11
10 +
01
```

Which is quite wrong. We would be saying that 3 + 2 = 1!

To get the correct answer, we must use at least 3 bits at a time. The number of bits taken at a time is called the 'word length'.

Generally, both inputs and outputs in a digital circuit have the same number of bits, so we must add zeroes to the left-hand side of a number in each of the unused columns.

The sum

```
  11
  10 +
 101
```

would be written as

```
011
010 +
101
```

If we did the same sum using an 8-bit word length, it would look like this:

```
00000011
00000010 +
00000101
```

The extra zeroes that are sometimes added are called 'leading zeroes'.

From the maths point of view, these leading zeroes appear as if we were making life difficult for no reason. It depends on whether we are looking at it as a sum to do, or as a digital design problem.

If we were doing it with a pen and paper we would immediately ignore all the leading zeroes, since we could always 'assume' they were there if the need ever arises.

If a digital circuit were to be built to perform the same addition, it would be quite unable to assume anything. If it needs to provide an 8-bit answer it must work in 8 bits throughout, and so we must have eight wires connected to the circuit.

If we were to add two binary numbers, the word length must be at least 1 bit longer than the longer of the two binary numbers to allow for the possibility of a final carry.

**Example using leading zeroes**

Here's the sum: add $11011_2 + 110_2$.

The first step is to decide on the number of bits to be used. The possibility of a carry occurring in the left-hand column will result in the maximum number of bits being one greater than the number of bits in the longer number. This is true when we are adding ANY two binary numbers.

In the example above, this rule would suggest that we work in 6-bit numbers so we add leading zeroes to bring both numbers up to 6 bits.

011011
000110 +

Then we can add in the normal way to give:

011011
000110 +
100001

Here's one for you to try.

**Example**

Using 8-bit numbers throughout, add $1111_2$ and $111_2$. The answer is $00010110_2$. If you run into trouble, the working is shown below.

Answer:

Step 1: Add leading zeroes until each number has 8 bits. The sum is now

00001111 + 00000111

Step 2: Lay them out making sure that the columns are properly aligned.

```
00001111
00000111
--------
```

Step 3: Add the columns, being careful not to forget the 'carries'.

```
00001111
00000111
--------
00010110 answer
--------
    1111  carries
```

## Subtraction of binary numbers

We can all subtract denary numbers and we can use the same method for subtracting binary numbers. However, when we ask a digital circuit to subtract binary numbers, we use a quite different method.

### Why change a method that works?

The advantage of doing this is that it allows us to re-use the adding circuit that we have already designed. There is a slight modification necessary, but it is very slight and, more importantly, extremely cheap to produce. Figure 3.1 shows the system used for adding two binary numbers and, by comparing it with the subtraction system in Figure 3.2, we can see that the only difference is that an extra circuit has been added. By switching this extra part in or out, we can instantly change from adding to subtracting.

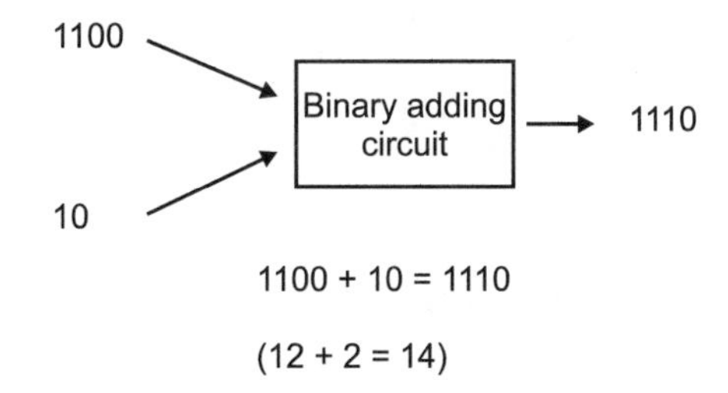


**Figure 3.1**
Binary addition

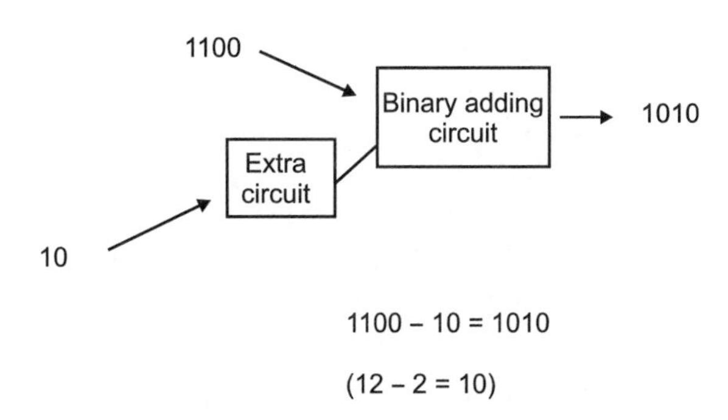


**Figure 3.2**
Binary subtraction using the adding circuit

### So how does this new system work?

If we take the number 12 and subtract 2, we get the answer 10. Fine. If we start with the number 12 and then add MINUS 2, we still get the same answer.

In figures, we can write this as $12 - 2 = 10$ or we can say $12 + (?) = 10$. The value of (?) must be $-2$ so that we can rewrite the sum as $12 + (-2) = 10$.

### So what?

So we are now adding two numbers, (12) and (−2). And as we are adding numbers, we can use the binary adder circuit that we have already built.

### Is there a snag?

Unfortunately, yes. The snag is how to load the number (−2). The only inputs to digital circuits are two voltages, one for the value 0 and the other for 1. We do not have an input voltage level that means 'minus'.

### The cure

We can get around this problem by replacing the −2 with another number that CAN be added to give the right answer. It may seem amazing, but we can find a positive number that can replace −2 to give the correct answer. Most people at this stage feel somewhere between confused and utterly confused, but just stay with it – things will improve. This magic number is called the complement of the number.

The calculation can now be written as 12 + (complement of 2) = 10.

### How do we find this magic number?

First, make sure that each number has the same number of bits by adding leading zeroes as we did for addition. In our example, the number $10_2$ is changed to $0010_2$.

The calculation now becomes as 1010 + (complement of 0010) = 10.

Finding the complement of 0010 is done in two stages. First we invert all the ones and zeroes. That means that all ones are changed to zeroes, and all zeroes are changed to ones. The result is called the 'one's complement' (don't worry about the name). In our example, the number 0010 becomes 1101.

We now add 1 to the one's complement so 1101 + 1 = 1110. This new number is called the 'two's complement'.

It's a bit confusing having two types of complement. We have to know, assume or guess which number is being referred to when books and designs talk about 'the complement'. If no clue is given, then assume that they are referring to the two's complement.

## Two's complement addition

Step 1: Add leading zeroes to make sure that both numbers have the same number of bits.

Step 2: Invert all the bits in the number we are subtracting. This means that we must change all zeroes into ones, and all ones into zeroes.

Step 3: Add 1 to the result of Step 2.

Step 4: Add the first number and the complement of the second number.

Step 5: Ignore the last carry bit.

### Example

Subtract the number 100 from 1011 using two's complement addition (or using complementary arithmetic).

Step 1: Here is the sum to be done: 1011 – 100.

Add a leading zero to the 100 to make them both up to 4 bi4s. This gives 1011 – 0100.

Step 2: Now invert 0100 to give 1011 (this is the one's complement of 0100).

Step 3: Add 1 to the number being taken away to make it the two's complement. This becomes 1011 + 1 = 1100 and rewrites the sum to an adding process.

It is now 1011 + (two's complement of 0100).

Step 4: Add these two numbers.

```
 1011
 1100 +
 ----
```

Starting with the right-hand column as usual, 1 + 0 = 1 with no carry.

```
 1011
 1100 +
 ----
    1
```

Next column, 1 + 0 = 1, no carry.

```
 1011
 1100 +
 ----
   11
```

Next column, 0 + 1 = 1, no carry.

```
 1011
 1100 +
 ----
  111
```

Left-hand column, 1 + 1 = 0, carry = 1.

```
  1011
  1100 +
  ----
  0111
1 ----  carry
```

Final addition by adding the carry.

```
   1011
   1100 +
  -----
  10111
1 -----  carry
```

Step 5: Since we designed our digital circuit to work with 4 bits, only four wires will be connected to the output to carry the answer. Any further information will simply be ignored by the circuit.

Our final answer to this subtraction was 10111. This is a 5-bit number, so we arrange for the left-hand bit to be the one ignored.

The final answer to the subtraction is 0111.

### A quick way to find the two's complement of a number

If you need to add any leading zeroes, do this first, then simply start from the left-hand end and change each 0 bit to a 1 and each 1 bit to a 0. Continue doing this until you come to the last '1'. Don't change this bit, and don't change any bit after it.

**Example**

Find the two's complement of 001011000.

Change each bit until the last '1'. Starting from the left-hand end, we would change the 00101 into 11010, then copy in the rest of the number without making any more changes. The result is 110101000. See Figure 3.3.

**REMEMBER** – Add the leading zeroes before finding the two's complement.

## You may want to skip this bit

The term 'one's' complement is derived from the way it can be generated. In this book we found it by inverting each bit in the number. This method was chosen because it is the way it is done in real digital designs. The alternative method is to subtract the number from a series of ones.

**Figure 3.3**

A quick way of finding the two's complement of a binary number

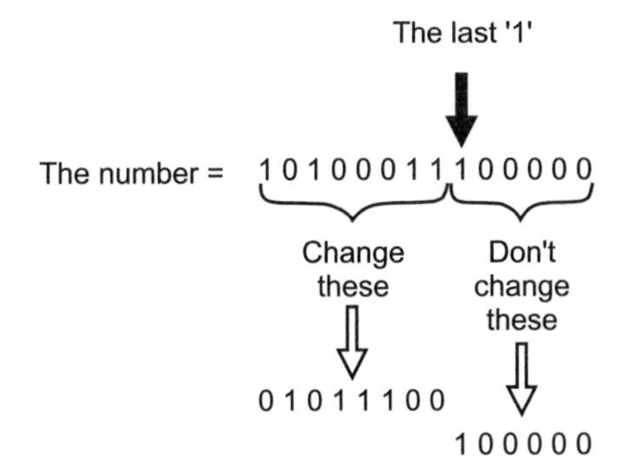


We started with 0010 and inverted it to 1101 to find the one's complement. The other way is to write down a line of ones with the same number of bits as are in the number and then subtract them.

Like this:

1 1 1 1
0 0 1 0
1 1 0 1 This is also the one's complement, so both methods work OK.

## Just for fun

This method of using complements can actually be used for subtraction using 'normal' denary numbers. It's very unlikely that anyone would find it better or easier than our other methods, but just for the fun of it we will run through an example.

We will subtract 50 from 200. Each step is just a modified version of the subtraction by two's complement in binary subtraction.

Step 1: Add leading zeroes to make sure that both numbers have the same number of digits.

Subtracting 50 from 200 is now written as:

200 – 050

Step 2: Subtract all the digits in the number we are subtracting, from a series of nines.

999 – 050 = 949

Step 3: Add 1 to the result.

949 + 1 = 950

Step 4: Add the result to the other number in the sum.

950 + 200 = 1150

Step 5: We are working in three digits so the extra digit that has occurred in the left-hand column is not required, therefore we delete it.

So 1150 becomes 150, which is the correct answer. Wow. Try one yourself.

## Quiz time 3

In each case, choose the best option.

**1 The result of adding $111_2$ to $1010_2$ would be:**

(a) $110011_2$
(b) $11010_2$
(c) $000110_2$
(d) $10001_2$

**2 The binary number 00101000 includes:**

(a) one leading zero.
(b) two leading zeroes.
(c) three leading zeroes.
(d) eight leading zeroes.

**3 The two's complement of an odd number always:**

(a) ends with a one.
(b) ends with a zero.
(c) starts with a zero.
(d) starts with a one.

**4 The result of $1101110101_2 - 1111_2$ is:**

(a) $870_{10}$
(b) $678_{10}$
(c) $101110101_2$
(d) $1101110111_2$

**5 The two's complement of 101011001110 is:**

(a) 010100110110
(b) 101011001111
(c) 100100110001
(d) 010100110010

# 4

# Hex to its friends

## The main problem with binary is us

We make fewer mistakes when we are interested, but unfortunately binary is not very interesting. In fact it is not interesting at all. Worse than that, the numbers are very long. The speed of light in metres per second can be written in denary as $299792459_{10}$, or in binary as $10001110111100111100001001011_2$. Which of these two numbers would we rather write down if we had to do it a hundred times and guarantee total accuracy? A modern digital circuit can handle binary numbers like this at the rate of several million a second.

We can make life easier by splitting the numbers up into groups to provide bite-sized portions. This improves both the denary as well as the binary. In denary we use groups of three, starting from the right-hand side, changing 299792459 into the much easier 299 792 459. In the binary world we split the numbers into groups of four, again starting from the right-hand end, so 10001110111100111100001001011 is written and read as 1 0001 1101 1110 0111 1000 0100 1011.

Notice also that the binary number is much longer than the equivalent denary number. In our example above, 29 bits has been compacted down to nine digits.

In an attempt to overcome these two shortcomings with binary (the number of digits and the ease of making mistakes), we sometimes use an alternative system called hexadecimal – or hex to its friends.

## Counting in hex

Binary uses two different bits.
Denary uses ten different digits.
Hex uses 16 different digits.

Hexadecimal counts from 0 to 15, which means 16 different digits, and so has a base of 16.

To keep it as simple as possible we use 0123456789 as in denary for the first ten digits then, for the last six, we use the letters ABCDEF or abcdef.

The hex system starts as:

| hex | denary |
|---|---|
| 0 | 0 |
| 1 | 1 |
| 2 | 2 |
| 3 | 3 |
| 4 | 4 |
| 5 | 5 |
| 6 | 6 |
| 7 | 7 |
| 8 | 8 |
| 9 | 9 |
| A | 10 |
| B | 11 |
| C | 12 |
| D | 13 |
| E | 14 |
| F | 15 |

When we run out of digits, we put a 1 in the second column and reset the first column to zero just as we always do.

So the count will continue:

| | |
|---|---|
| 10 | 16 |
| 11 | 17 |
| 12 | 18 |
| 13 | 19 |
| 14 | 20 |
| 15 | 21 |
| 16 | 22 |
| 17 | 23 |
| 18 | 24 |
| 19 | 25 |
| 1A | 26 |
| 1B | 27 |
| 1C | 28 |

| | |
|---|---|
| 1D | 29 |
| 1E | 30 |
| 1F | 31 |
| 20 | 32 |

and so on.

It takes a moment or two to get used to the idea of having numbers that include letters, but it soon passes.

As with binary, we must be careful to include the base whenever necessary to avoid confusion. The base is usually written as H, though h or 16 would still be acceptable.

'One eight' in hex is equal to 24 in denary. Notice how I avoided quoting the hex number as eighteen. Eighteen is a denary number, and does not exist in hex. If you read it in this manner it reinforces the fact that it is not a denary value.

In various books and datasheets, we meet hex numbers written in slightly different forms.

Here are the main options in order of popularity:

$24_{10}$ = 16H or $16_H$ or 16h or $16_h$ or $16_{16}$.

## The advantages of hex

1. It is very compact. Using a base of 16 means that the number of digits used to represent a given number is usually less than in binary or denary.
2. It is quick and easy to convert between hex and binary, and fairly easy to go between hex and denary.

## Converting denary to hex

The process follows the same pattern as we saw in the denary to binary conversion.

### Method

1. Write down the denary number.
2. Divide it by $16_{10}$, put the whole number part of the answer underneath and the remainder in the column to the right.
3. Keep going until the number being divided reaches zero.
4. Read the answer from the bottom to top of the remainders column.

REMEMBER TO WRITE THE REMAINDERS IN HEX.

**Example**

Convert the denary number 3077 to hex.

Step 1: Write down the number to be converted.

```
3077
```

Step 2: Divide by 16. You will need a calculator. The answer is 192.3125. The 192 can be placed under the number being converted.

```
3077
 192
```

The decimal part of the answer, 0.3125, is actually 0.3125 of 16. Multiply 0.3125 by 16 and the result is 5. Write the 5 down in a separate column to the right.

When completed, this step looks like:

```
3077
 192      5
```

Step 3: Repeat the process by dividing the 192 by 16 to give 12. There is no remainder, so we can just enter the result as 12 with a zero in the remainder column.

```
3077
 192      5
  12      0
```

Step 4: Now, 12 is less than 16, so the answer is going to be 0 with a remainder of 12. Great care here – the 12 is a decimal number which is C in hex, so the number in the remainder column is written as C.

Enter the values in the normal columns to give:

```
3077
 192      5
  12      0
   0      C
```

Step 5: Read the hex number from the bottom upwards: C05H (remember that the 'H' just means a hex number).

```
3077
 192      5   ↑    3077₁₀ = C05H
  12      0   |
   0      C
```

$3077_{10}$ = C05H

And one for you to try. The answer follows.

**Example**

Convert 44 $256_{10}$ into hex.

Answer:

| | | | |
|---|---|---|---|
| 44256 | | | |
| 2766 | 0 | ↑ | = ACE0H |
| 172 | E | | |
| 10 | C | | |
| 0 | A | | |

Remember to add the H for hex on the end of the number, particularly in cases where the hex number doesn't contain any letters.

## From hex to denary

To do this we can use a similar method to the one we used to change binary to denary, except that in this case each column is 16 times larger than the one to the right.

### Example

Convert B02F9H to denary.

Step 1: The column values are:

| $16^4$ | $16^3$ | $16^2$ | $16^1$ | $16^0$ |
|---|---|---|---|---|
| 65536 | 4096 | 256 | 16 | 1 |

Notice how $(\text{anything})^0 = 1$, so $10^0 = 2^0 = 16^0 = 1$.

Step 2: Simply enter the hex number using the columns.

| 65536 | 4096 | 256 | 16 | 1 |
|---|---|---|---|---|
| B | 0 | 2 | F | 9 |

Step 3: The value of the right-hand column is $9 \times 1 = 9$.
The next column is $F \times 16$, which is $15 \times 16 = 240$.
The third column is $2 \times 256 = 512$.
The fourth column is zero.
The last column is $B \times 65536$ or $11_{10} \times 65536 = 720896$.

| 65536 | 4096 | 256 | 16 | 1 |
|---|---|---|---|---|
| 720896 | 0 | 512 | 240 | 9 |

Step 4: Add up all the denary column values:

$720896 + 512 + 240 + 9 = 721657$

## Method

Step 1: Write down the column values using a calculator. Starting on with $16^0$ (= 1) on the right-hand side and increase by 16 times in each column towards the left.

Step 2: Enter the hex numbers in the appropriate columns.

Step 3: Use a calculator to find the denary values of each column.

Step 4: Add all the column totals to obtain the final denary equivalent.

Try this one.

**Example**

Convert C0E4H to denary.

Answer:

| $16^3$ | $16^2$ | $16^1$ | $16^0$ | column values |
|---|---|---|---|---|
| 4096 | 256 | 16 | 1 | column values |
| C | 0 | E | 4 | Hex numbers |

First column: $4 \times 1 = 4$
Second column: $E \times 16 = 14 \times 16 = 224$
Third column: zero
Fourth column: $C \times 4096 = 12 \times 4096 = 49152$

Total = $49152 + 224 + 4 = 49380_{10}$

## Binary to hex conversion

If we have a 4-bit binary number, its lowest value is $0000_2$ or zero. Its highest value is $1111_2$, which converts to $8 + 4 + 2 + 1 = 15$ in denary.

This means that any group of 4 bits can be translated directly into a single hex digit. Just put 8, 4, 2, and 1 over the group of bits, and add up the values wherever a 1 appears in the binary group.

**Example**

Convert $11110100000 11_2$ to hex.

Step 1: Starting from the right-hand end, chop the binary number into groups of four.

11 1101 0000 0011

Step 2: Convert each group of 4 bits to a hex number. The right-hand group is 0011, so this will convert to:

8 4 2 1 column headers
0 0 1 1 binary number
0 0 2 1 column values

The total will then be $0 + 0 + 2 + 1 = 3_{10}$ also 3 in hex.

The right-hand side binary group can now be replaced by the hex value 3.

| 11 | 1101 | 0000 | 0011 |
|---|---|---|---|
| | | | 3 |

Step 3: The second group can be treated in the same manner:

The bits are all zeroes so this one is easy, the answer is zero in hex or OH. Adding the zero under the second group gives:

| 11 | 1101 | 0000 | 0011 |
|---|---|---|---|
| | | 0 | 3 |

Step 4: The next group is 1101, which translates into $8+4+1=13$. Converting the $13_{10}$ into hex gives D. The result so far will be:

| 11 | 1101 | 0000 | 0011 |
|---|---|---|---|
| | D | 0 | 3 |

Step 5: The last group is incomplete, so only the column headings of 2 and 1 are used. If you think it helpful, you could start by adding leading zeroes to the original number to keep it in blocks of 4 bits.

In this case, the result from $11_{10}$ or from $0011_{10}$ would be $3_{10}$ and 3H.

This gives a final result of:

| 11 | 1101 | 0000 | 0011 |
|---|---|---|---|
| 3 | D | 0 | 3 |

So $11110101000011_2$ = 3D03H.

**Note**: Always remember to start chopping from the right-hand side.

## Method

1. Write down the binary number.
2. Chop it into groups of 4 bits starting from the right-hand end.
3. Convert each block of 4 bits into a denary number and then into a hex number.

### Example

Here is one to try. As usual, the solution follows.

Convert the number $1001010110111_2$ to hex.

Answer:

Step 1: Write down the binary number.

$1001010110111_2$

Step 2: Chop into groups of four starting from the right-hand side.

1 0010 1011 0111

Step 3: Left-hand group = 0111 = 4 + 2 + 1 = $7_{10}$ = 7H
Second group = 1011 = 8 + 2 + 1 = $11_{10}$ = BH
Third group = 0010 = $2_{10}$ = 2H
The last group is just a single 1, so it becomes 1H.

1/0010/1011/0111
1 2 B 7

Final result: $1001010110111_2$ = 12B7H.

## Changing hex into binary

This is just the reverse of the last process. Simply take each hex number and express it as a four bit binary number.

As we saw in the last section, a 4-bit number has column header values of 8, 4, 2 and 1, so conversion is just a matter of using these values to build up the required value. All columns used are given a value of 1 in binary, and all unused columns are left as zero.

When we are converting small numbers like 3H we must remember to add zeroes on the left-hand side to make sure that each hex digit becomes a group of 4 bits.

Imagine that we would like to convert 5H to binary. Looking at the column header values of 8, 4, 2 and 1, how can we make the value 5? The answer is to add a 4 and a 1. Taking each column in turn: we do not need to use an 8 so the first column is a 0. We do want a 4 so this is selected by putting a 1 in this column. We don't need a 2, so make this column 0 and finally put a 1 in the last column to select the value of 1. The number 5H is now converted to $0101_2$.

All values between 0 and FH are converted in a similar way.

**Example**

Convert 2F60H to binary.

Step 1: Write the whole hex number out with enough space to be able to put the binary figures underneath.

| 2 | F | 6 | 0 |
|---|---|---|---|

Step 2: Put the column header values below each hex digit.

| 2 | F | 6 | 0 |
|---|---|---|---|
| 8421 | 8421 | 8421 | 8421 |

Step 3: The hex 0 does not require any values from the columns but do remember to use a 4-bit replacement number, so convert it to 0000.

| 2 | F | 6 | 0 |
|---|---|---|---|
| 8421 | 8421 | 8421 | 8421 |
| | | | 0000 |

Step 4: Now do the same for the next column. The hex number is 6, which is made of 4 + 2, which are the middle two columns. This will result in the binary group $0110_2$.

| 2 | F | 6 | 0 |
|---|---|---|---|
| 8421 | 8421 | 8421 | 8421 |
| | | 0110 | 0000 |

Step 5: Since 8 + 4 + 2 + 1 = 15, the hex F will become $1111_2$.

| 2 | F | 6 | 0 |
|---|---|---|---|
| 8421 | 8421 | 8421 | 8421 |
| | 1111 | 0110 | 0000 |

Step 6: Finally the last digit is 2, and since this corresponds to the value of the second column it will be written as $0010_2$.

| 2 | F | 6 | 0 |
|---|---|---|---|
| 8421 | 8421 | 8421 | 8421 |
| 0010 | 1111 | 0110 | 0000 |

The final result is 2F60H = $0010111101100000_2$.

But do we include the two leading zeroes? The answer may be 'yes' or 'no'. It depends on what we are doing. If we were doing a calculation then we would want the numerical value and the two leading zeroes wouldn't matter. If, however, we were finding information as part of the design of a digital circuit, we would have 16 wires and we would need to know what voltage to apply to each of the wires. In this case the leading zeroes would be needed.

## Method

Step 1: Write down the hex number but make it well spaced.

Step 2: Using the column header values of 8, 4, 2 and 1, convert each hex number to a 4-bit binary number.

Step 3: Add leading zeroes to ensure that every hex digit is represented by 4 bits.

Here is an example for you to try. As usual, the answer is shown.

### Example

Convert 1375DH to binary.

Answer:

Step 1:

| 1 | 3 | 7 | 5 | D |
|---|---|---|---|---|
| 8421 | 8421 | 8421 | 8421 | 8421 |

Step 2:

0001 0011 0111 0101 1101

So 1375DH = $00010011011101011101_2$.

## Using stepping stones

Some conversions are easier than others.
Between binary and hex they are easy.
Between binary and denary they are not too bad.
But between hex and denary they are difficult and nearly always need a calculator.

Sometimes a longer but simpler route is a good idea. To convert from hex to denary, it may be easier to convert the hex to binary and then the resulting binary to denary. Similarly, from denary to hex, we could go from denary to binary and then from binary to hex (see Figure 4.1).

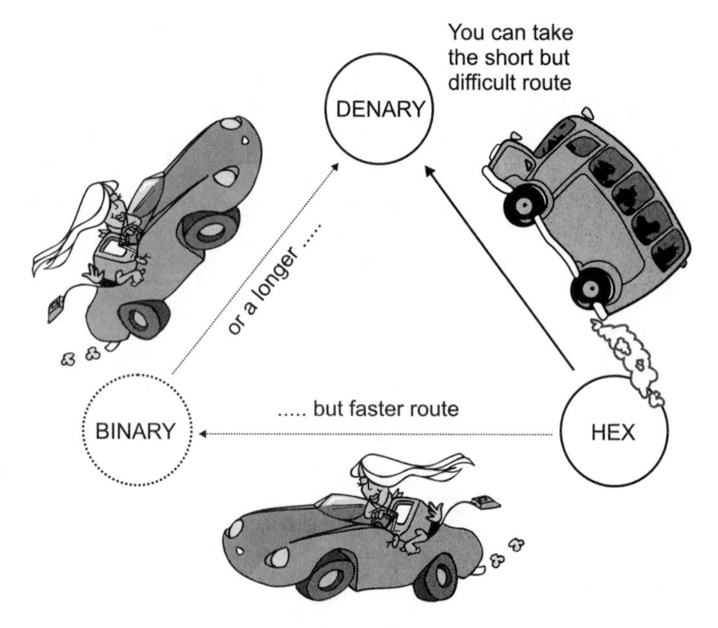


**Figure 4.1**
There is more than one way to convert denary to hex

## Obsolete octal – probably not worth reading

Octal is another number system, which has almost disappeared. It uses only eight digits.

The count proceeds:

0
1

2
3
4
5
6
7 there is no eighth digit, so reset the count to 0 and put a 1 in the next column
10
12
13
14
15
16
17 now go straight to 20
20 (two zero – remember, it isn't 'twenty')
etc.

As no letters are used, it may take a little while before we realize that it is octal rather than denary.

Conversions follow the same pattern as we have seen for hex.

Octal to denary: the column heading values are $8^4$, $8^3$, $8^2$, $8^1$, $8^0$.

Denary to octal: divide by eight and write down the remainders, then read remainders from the bottom upwards. Use the subscript 8 to indicate an octal number, e.g. $64_{10} = 100_8$.

Octal to binary: write each octal digit down as a *three* digit binary group.

Binary to octal: start from the right-hand side and chop the binary numbers into groups of three, then evaluate each group.

That's about all it is worth doing on octal. There are some digital circuits that are associated with computers that still use octal, but they are very rare.

## Quiz time 4

In each case, choose the best option.

**1 Which of these represents the largest number?**

(a) $1000_8$
(b) $1000_{10}$
(c) 1000H
(d) $1000_2$

**2 The number 11D02H is equal to:**

(a) $72962_{10}$
(b) $4560_{10}$
(c) $79262_{10}$
(d) $7426_{10}$

**3 The highest digit in the hexadecimal system is:**

(a) 15
(b) H
(c) 16
(d) F

**4 Which of these numbers is the same as $101001010111_2$?**

(a) $10010111_{10}$
(b) $A57_{16}$
(c) 1057H
(d) 75Ah

**5 The number of digits in a hex number is always:**

(a) more than the number of digits in the equivalent binary number.
(b) more than its radix.
(c) less than or equal to the number of digits in the equivalent denary number.
(d) more than the number of digits in the equivalent denary number.

# 5

# Just logic

In which of the circuits in Figure 5.1 will the light be on?

**Figure 5.1**
Which lamps are ON?

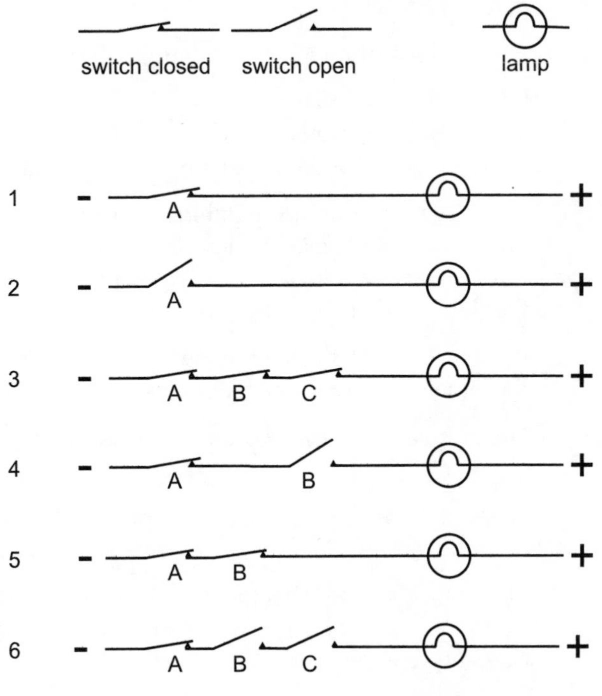

switch closed switch open lamp

1 - A +

2 - A +

3 - A B C +

4 - A B +

5 - A B +

6 - A B C +

Hopefully, you chose numbers 1, 3 and 5. In all three of these cases, the important factor was that all the switches were closed.

Taking circuit 5, we can say that switch A AND switch B had to be closed before the light came on.

Digital circuits are composed almost exclusively of groups of switches. It would not be at all unreasonable for a modern digital circuit to contain a million switches. For this reason, digital circuits are often called 'switching circuits'.

The group of switches that controls a circuit is given the curious name of a 'gate'.

As we have seen, in circuit 5 the light is ON, provided switches A AND B are both ON. We could say that switch A and switch B together form an AND gate.

## Making it difficult

It's never long before we start using some technical terms, but let's introduce them slowly.

The gate in circuit 5 had two switches, both of which must be closed if the light is to come on. We call the states of the two switches 'input conditions' or just 'inputs' and, since input A AND input B must both be correct for the light to come on, we call this gate a 2-input AND gate.

Have a look at circuit 3 in Figure 5.1. How would you describe this gate?

Yes, it is a 3-input AND gate. Notice how we use capital letters for the AND to avoid awkward sentences like '. . . and an and gate . . .'.

The 'states' or 'input conditions' of gates are normally described by numbers, and we usually describe a closed switch as a logic 1 and an open switch as a logic 0. Having done this, we can say that to make the light be ON we must have a logic 1 state on each of the inputs.

If we now define the light being ON as a logic 1 state, we can describe the situation in two different ways.

First, in plain English – when all the switches are ON, the light comes ON.

Second, the technical version – when all the inputs are at logic 1, the output is at logic 1.

The only advantage that the technical version has is that it can be used in a more general sense to mean 'when all the input conditions are met, the required outcome will occur'.

## Truth tables

A truth table is a simple list of all the possible input states and all the corresponding outputs.

With two switches, as in circuit 5, we have four possible situations in the circuits. These are shown in Figure 5.2.

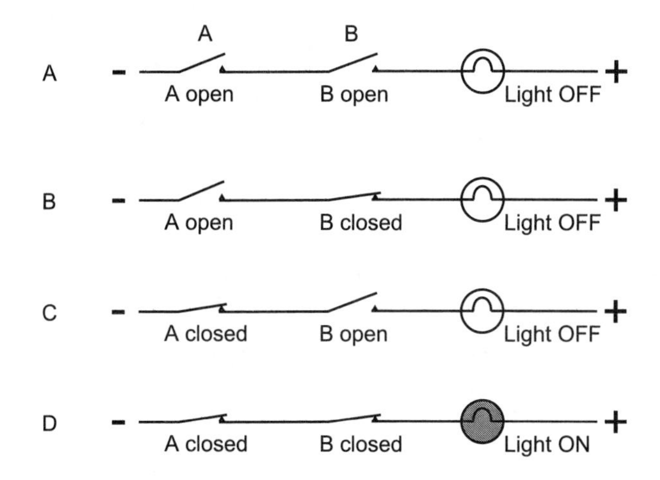


**Figure 5.2**
The four possible situations

Using the conventions:

Switch OFF = logic 0
Switch ON = logic 1
Light OFF = logic 0
Light ON = logic 1

we can describe each of the four situations in the following way.

Situation A could be described as A = 0, B = 0, light = 0, because both switches are OFF and the light is also OFF.

Situation B has switch A OFF and switch B ON, but the light is still OFF. This can be described as A = 0, B = 1, light = 0.

Situation C is very similar, but the switches are the opposite way around. So A = 1, B = 0, light = 0.

Situation D has both switches ON and hence the light is ON. This state is written as A = 1, B = 1, light = 1.

Rather than writing all the possible states in sentences as above, we just produce a table that lists all the possibilities, like this:

| Switch A | Switch B | Light |
|---|---|---|
| 0 | 0 | 0 |
| 0 | 1 | 0 |
| 1 | 0 | 0 |
| 1 | 1 | 1 |

This table is nice and easy to read, and gives all the relevant information about the operation of the circuit.

We usually make the table even more compact by abbreviating the column headings. The inputs could just be called A and B and the output just L (L for light):

| A | B | L |
|---|---|---|
| 0 | 0 | 0 |
| 0 | 1 | 0 |
| 1 | 0 | 0 |
| 1 | 1 | 1 |

A truth table is therefore just a compact method of describing all the possible conditions that can be met in a circuit.

Notice how the input possibilities are listed in binary order as we count up: 00, 01, 10, 11. It is not essential to do it this way, but it is an easy way to make sure that we don't miss any of the possible input states.

## A 3-input AND gate

A 3-input AND gate is very much the same as the previous example, except that there are three inputs that control the output.

The truth table would have three columns for the inputs and one for the output state. The three input columns could be called A, B and C, or anything else that we may think helpful. The three columns would just count up in binary, and the output would be OFF or at a 0 state until, on the last line, the three inputs would all be at 1 states and the output would switch ON or go to a 1 state.

Here it is:

| A | B | C | L |
|---|---|---|---|
| 0 | 0 | 0 | 0 |
| 0 | 0 | 1 | 0 |
| 0 | 1 | 0 | 0 |
| 0 | 1 | 1 | 0 |
| 1 | 0 | 0 | 0 |
| 1 | 0 | 1 | 0 |
| 1 | 1 | 0 | 0 |
| 1 | 1 | 1 | 1 |

With a 2-input gate, we had a total of $2^2 = 4$ rows in the truth table. In this 3-input gate, the number of rows has increased to $2^3 = 8$ rows. A 10-input gate would have $2^{10} = 1024$ rows.

## Circuit symbols

### The big match: the US Military Specification versus the rest of the world

The US Military Specification devised a series of distinctive shapes to represent digital gates and other circuits. These designs were happily accepted almost universally because they were quickly and easily recognized on a circuit diagram. As digital circuits increased in complexity and dealt with hundreds of gates at a time, a more compact system of symbols was needed.

### The authorities versus the people

In the 1980s the national and international standards authorities, one by one, decreed that we should adopt a series of new symbols, which were all basically rectangular.

By this time the US military symbols were well established and were (and are) very popular, and the change over was exceedingly slow. We were clearly dragging our feet and, to hasten the process, all UK schools, colleges and exam bodies were 'encouraged' to change so the change would eventually work through into industry.

### At the present time

We have a choice:

1 The American ANSI-IEEE-91–1991 version 2A and the ANSI Y32 employ all rectangles.
2 The European standard, called Euronorm 60717, which has been incorporated into British Standard 3939; this also employs all rectangles.
3 The International IEC 617–2 standard, which has also used the rectangles.

So everything is pretty well standardized on the same set of rectangular symbols. Or it would be except for the 'distinctive shape' symbols that refuse to die – so much so that the standards still allow the use of the distinctive shape symbols.

In the UK many of the examination bodies have now reverted to the use of the 'old' US military shapes on examination papers, and every digital engineer is brought up on the shapes in preference to the rectangles.

### Why all this difficulty?

The heart of the matter is that the rectangles and distinctive shapes are seen to be alternative, competing systems, whereas in reality they are complementary.

A small digital circuit using the distinctive shapes system is very much easier to read. When we are first introduced to digital circuits we see individual gates and small circuits for which the 'US Mil. Spec.' is ideal.

When we move on to larger circuits employing dozens or hundreds of gates and more complex integrated circuits, the compact nature of the rectangles is well suited. At this stage, of course, we have to change systems, and we hate it. Whether it is feet into metres, Fahrenheit into Celsius or changes of currency, our instinctive reaction is to dig our heels in and fight to carry on the way we have always done. This is doubly the case when we can point to something that is clearly better about the old system.

### What are we going to use?

As new circuit elements are introduced, both the US Mil. Spec. 'shapes' and the new 'rectangular' versions will be shown side by side. In subsequent diagrams, only the shapes will be used. Hopefully this will prove a gentle introduction to both systems.

### Using the symbols of either system

When we have a choice, we prefer to have the symbol drawn horizontally with the inputs to the left and the output on the right. In this way, we can 'read' the diagrams from left to right.

## An AND gate

A 2-input AND gate is shown in Figure 5.3, together with its truth table.

## The symbols

The US Military Specification symbol has a distinctive shape, whereas the International symbol is basically a rectangle. Notice in the rectangular symbol how the type of gate is always signified by a symbol placed in the top centre. This symbol shows that a single AND integrated circuit actually contains four separate AND gates as shown by the symbol being split into four by the horizontal lines.

The symbol for a 2-input AND gate does not say anything about how it was made, it just describes how it responds to the inputs applied. We built a perfectly valid 2-input gate from a couple of switches and used

**Figure 5.3**
2-input AND gate symbols and truth table

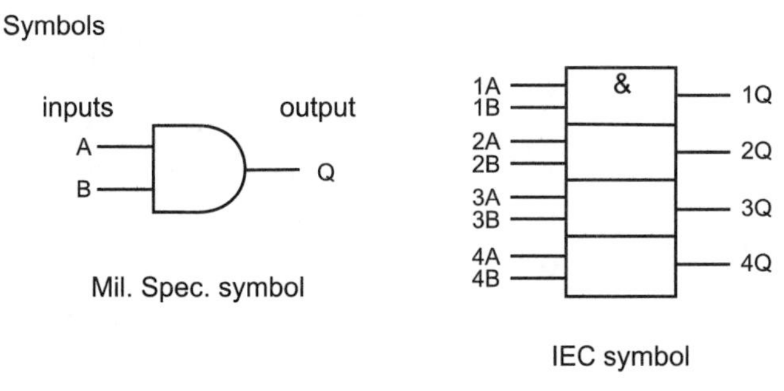


Truth table

| inputs | | output |
|---|---|---|
| A | B | Q |
| 0 | 0 | 0 |
| 0 | 1 | 0 |
| 1 | 0 | 0 |
| 1 | 1 | 1 |

The output is always 0 unless both inputs are 1

a lamp to indicate the output, but we could equally well have manufactured any other electronic circuit or device provided that it responds in the same way to applied inputs.

## The truth table

In our first truth tables, we used the letters A and B to represent the inputs and the letter L for the output. This was convenient since we built the circuit using a lamp for the output.

Letters A, B, C etc. are often used as the inputs to gates and, strangely enough, the letter Q is a popular choice for the output, but in neither case are these letters compulsory. We can, if we wish, use our own choice of letter.

## Boolean algebra

Boolean algebra was first invented in 1847 by an Englishman called George Boole to provide a way of calculating logical thought processes. Things like: 'all metals can melt' and 'all ice cream can melt'; so is it necessarily true that all ice creams are made from metal?

George got it wrong on his first attempt and had to re-invent it in 1854, adding a note requesting his readers to disregard his earlier paper as it was neither complete nor correct. Second time lucky. He was the first mathematician for 200 years to attempt the manipulation of logical statements by symbols and universal rules.

One of its first applications was to simplify the design of the switching mechanisms in early telephone exchanges.

Boolean algebra is a means of writing down the description of a gate or a combination of gates to avoid the necessity of drawing all the symbols. As we will see later, it also allows us to simplify the design of logic circuits.

## Ways of describing an AND gate

If we have a 2-input AND gate with inputs A and B and an output of Q, we could describe what it does in words: The output is logic 1 only when both inputs are at logic 1.

An alternative way is to say 'A AND B = Q'. This is a little quicker than the previous definition, and Boolean algebra takes it one step further by replacing the word AND by a dot.

So A AND B = Q becomes A.B = Q.

Even the Boolean expression is sometimes abbreviated further by leaving out the dot between the A and the B, so A.B = Q can be written as AB = Q. Whether we use the dot symbol is entirely optional. Both forms are read as 'A AND B equals Q'.

A 3-input gate could be written as A.B.C = Q or ABC = Q, and is read as 'A AND B AND C equals Q'.

So far, an AND gate can be described:

1 In words.
2 By Boolean algebra.
3 By circuit symbols.
4 By a truth table.

### Example

Use each of the above methods to describe a 3-input AND gate.

Answer:

In words: The output will be at a logic 1 if, and only if, all three of the inputs are at logic 1.

As a Boolean expression: A.B.C = Q or ABC = Q. Any other letters can be used if required.

The circuit diagram and truth table are shown in Figure 5.4.

**Figure 5.4**
The 3-input AND gate symbol and truth table

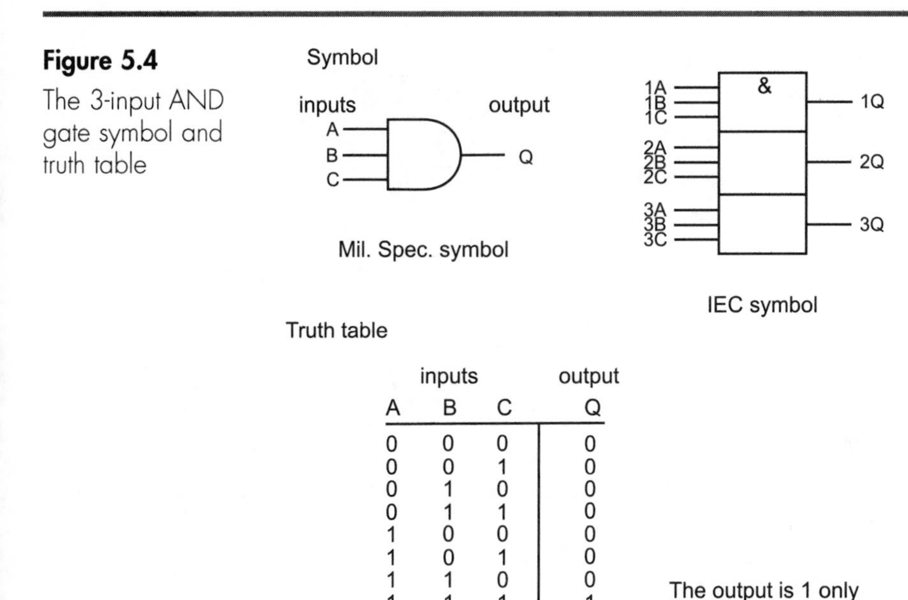


| inputs | | | output |
|---|---|---|---|
| A | B | C | Q |
| 0 | 0 | 0 | 0 |
| 0 | 0 | 1 | 0 |
| 0 | 1 | 0 | 0 |
| 0 | 1 | 1 | 0 |
| 1 | 0 | 0 | 0 |
| 1 | 0 | 1 | 0 |
| 1 | 1 | 0 | 0 |
| 1 | 1 | 1 | 1 |

## AND gates with many inputs

So far the circuit symbol has shown each input as a separate line, but the symbol often gets too crowded as the number of inputs increase. When this happens, we can change the symbol slightly as shown in Figure 5.5.

**Figure 5.5**
Some gates have many inputs

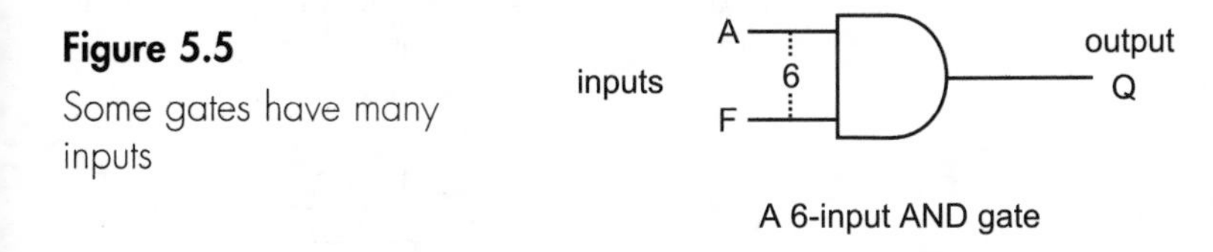


## The OR gate

Have a look at the circuits in Figure 5.6 and see which lamps will be lit.

The light will be ON in all cases except for the first one. Since the switches are connected in parallel, the light will come ON when switch A is closed or when switch B is closed, or when both are closed.

A circuit that behaves in this way is called an OR gate.

Just like the AND gate, it can be described in four ways.

**Figure 5.6**
Which lamps are ON?

## In words

The output of an OR gate is at logic 1 if any, or all of the inputs, are at logic 1.

## As a Boolean expression

We use a + sign to mean OR, so a 2-input OR gate could be written as $A+B=Q$. A 3-input OR gate could be written as $F+K+R=T$ or any other letters we care to use.

REMEMBER, + means OR. At first (or even second) glance, using a dot to mean AND and the + symbol to mean OR may seem curious. In Chapter 8 we will look at the reasoning behind it.

## By symbols and truth tables

These are shown in Figure 5.7. Notice the new shape used for the OR gate. This makes it easily recognized on a logic diagram. Gates with more inputs are catered for in the same way as AND gates.

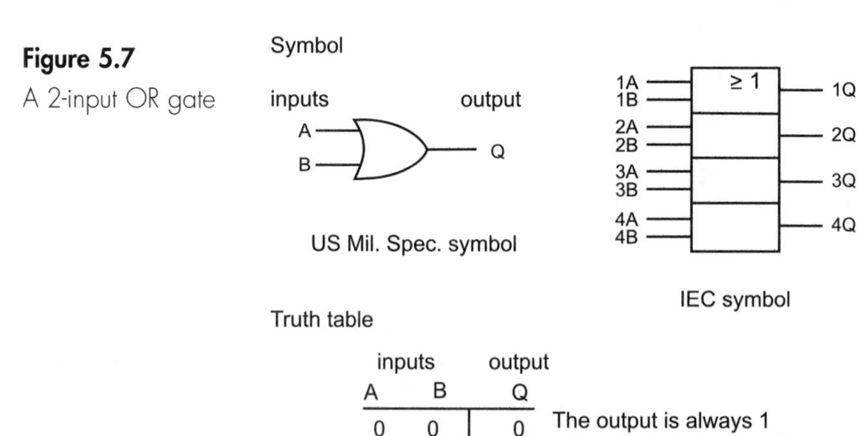


**Figure 5.7**
A 2-input OR gate

| inputs A | inputs B | output Q |
|---|---|---|
| 0 | 0 | 0 |
| 0 | 1 | 1 |
| 1 | 0 | 1 |
| 1 | 1 | 1 |

The output is always 1 unless both inputs are 0

## A quick English lesson

Read these two conversations and answer the question 'What does the word "or" mean?'

A friend says: 'Would you like something to eat or drink?'
You reply 'Yes please, a burger and a coffee'.

A colleague asks: 'Is it Monday or Tuesday today?'
You answer 'It's Tuesday'.

We are so familiar with English that we know without a moment's thought that the two 'or's have different meanings.

The first case is what we call an inclusive OR. What we really mean is 'Would you like something to eat or something to drink OR BOTH?

In the second case, we are using an exclusive OR. We could answer 'Monday' or 'Tuesday', but NOT BOTH Monday and Tuesday.

## Back to digital

Take a glance at the truth table for the OR gate in Figure 5.7. Is this behaving like an inclusive or an exclusive 'or'? It is using the inclusive meaning of the word 'or', since the last line of the table shows that the output is at a logic 1 when both of the inputs are at a logic 1.

## The exclusive-OR gate

In some design situations we would like to use the exclusive form to prevent two events occurring at the same time. Perhaps on a lift someone has pressed the 'up' and 'down' buttons at the same time.

Exclusive-OR is often abbreviated to XOR, EXOR or EOR.

Note: XOR gates only occur in 2-input versions.

## In words

The output of an exclusive-OR gate is a logic 1 state if either, but not both, of the inputs are at a logic 1. Another way of saying this is that the output goes to a logic 1 only if the two inputs are at different logic levels and this has resulted in its alternative name of the 'difference gate'.

## As a Boolean statement

We use the symbol $\oplus$ to mean 'exclusive-OR', so an XOR gate could be described as $A \oplus B = Q$.

The symbol appears as a modified OR gate and is shown in Figure 5.8.

**Figure 5.8**

The exclusive-OR, XOR, EXOR, EOR or difference gate

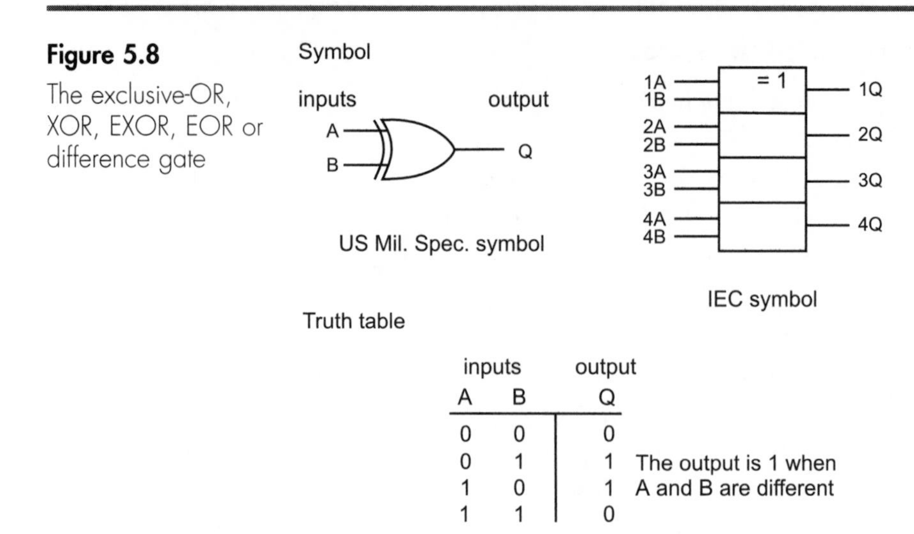


| inputs A | inputs B | output Q |
|---|---|---|
| 0 | 0 | 0 |
| 0 | 1 | 1 |
| 1 | 0 | 1 |
| 1 | 1 | 0 |

The output is 1 when A and B are different

## The NOT gate

This has got to be the simplest gate ever. It has only one input and one output, and a truth table with only two lines.

So what does it do? It just reverses the logic state.

This gate is also called an inverter since its function is to invert the logic state.

If we apply logic 1 at the input, the output becomes logic 0.
If we apply logic 0 at the input, the output becomes logic 1.

## In words

The output always has the opposite logic level to the input.

## The Boolean description

The Boolean symbol for a NOT is a line or 'bar' over the letter, so if the input to a NOT gate is written as A, then the inverted output would be written as $\overline{A}$.

The output to our example in Figure 5.9 would be written as $Q = \overline{A}$.

## The symbol and truth table

These are shown in Figure 5.9.

The inversion of the output signal is shown by a small circle in the US Mil. Spec. symbol and by a polarity indicator shown on the rectangular symbol.

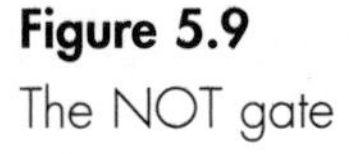

**Figure 5.9**
The NOT gate

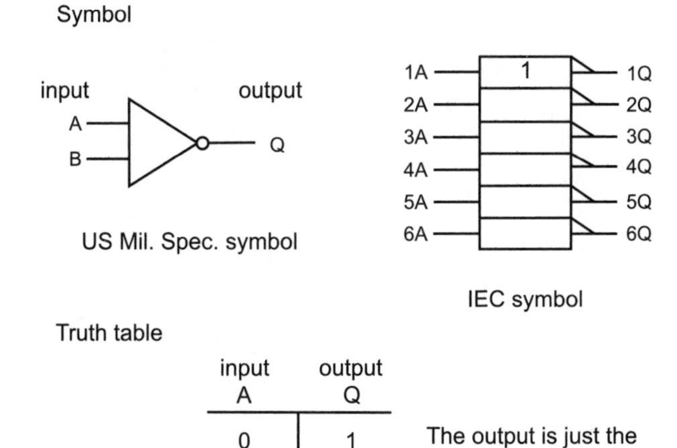


| input A | output Q |
|---|---|
| 0 | 1 |
| 1 | 0 |

If we apply an input A to an inverter, we would get $\overline{A}$ at the output. The term $\overline{A}$ would be read as 'NOT A'.

Likewise, if we applied this $\overline{A}$ as the input, then we would get the inverted version at the output; but how would we write it?

We could write this in two ways.

First, we could say that the effect of a NOT gate is to invert the input, whatever it is. As we know, we show an inversion by putting a line over the top of the input signal. So, if we have an input of $\overline{A}$ then the output can be written as $\overline{\overline{A}}$.

Alternatively, we could argue that if the logic state was changed and then changed back again, then it must be back to the original value. Therefore we could write the output simply as A as in Figure 5.10.

**Figure 5.10**
Outputs from a NOT gate

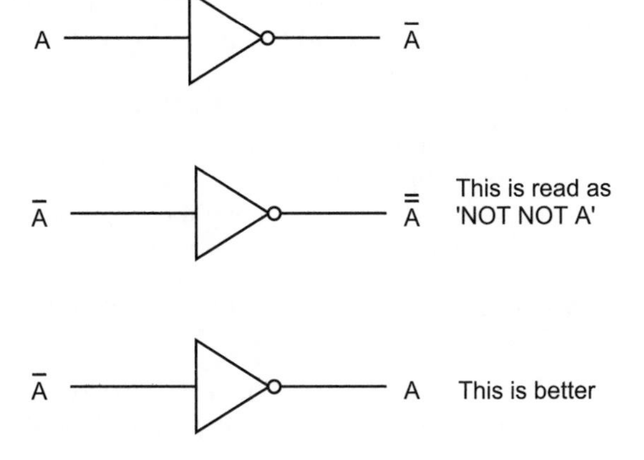

From this we can see that $\overline{\overline{A}} = A$ so whenever a double line is met, it can be simplified by removing both of the 'bars'. It is always a good idea to simplify multiple bars as much as possible as soon as they occur.

How would we write the output of the circuit in Figure 5.11?

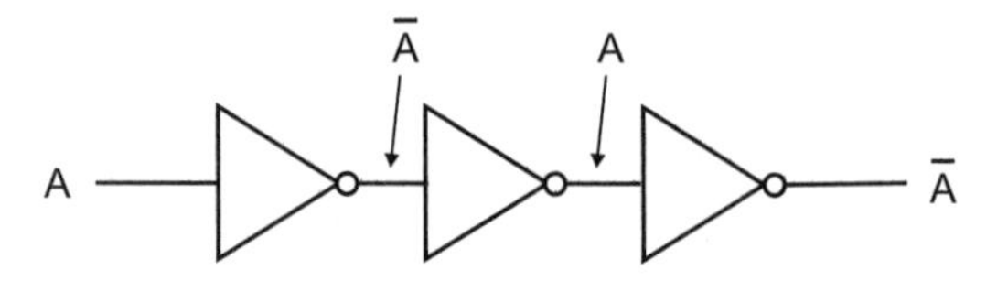


**Figure 5.11**
Always simplify as much as possible

We would write it as $\overline{A}$, the alternative $\overline{\overline{\overline{A}}}$ is correct but is not used since it would be unnecessarily clumsy. This is like using multiple negatives in English. We wouldn't say 'it is not not not raining today'.

It is interesting to see that the NOT gate actually produces the one's complement of the input, so a group of four NOT gates could be used to find the one's complement of a 4-bit number.

## So far . . .

So far we have looked at four gates – AND, OR, XOR and NOT. These are the only basic types of gate that occur. There are three others, but they are just combinations of these basic types. It often happens that when designing a real circuit we find that the output signal needs to be inverted. We could, of course, just add an inverter to reverse the logic level, but the problem occurs so often that we build a version of AND, OR and XOR gates that already have the inverters built in.

This has three advantages:

1 It saves cost – the price is much the same with or without the inverter, whereas a gate and an inverter bought separately would double the total cost.
2 It saves space – the built-in NOT gate does not increase the total size of the original gate at all.
3 It saves time – adding an internal NOT gate results in no additional time delay over and above the original gate, whereas an external NOT would double the total time.

Internal NOT gates are cheaper, smaller and faster, and therefore very popular.

Remember that the 'new' gates that follow are only different because the output logic states have been inverted. This inversion is shown by adding a small circle at the output like we had in the symbol for the NOT gate. Have a glance back to Figure 5.9.

## The NAND gate

NOT AND has been abbreviated to the word NAND.

## In words

The output from the NAND gate is a logic 1 unless both, or all, inputs are at logic 1.

## As a Boolean expression

The inversion by the internal NOT gate is shown by adding the line over the normal AND output, and is exactly the same result as we would have if we added a real NOT gate to an AND gate.

You will recall that a 3-input AND gate would have an output of Q=ABC or Q=A.B.C.

The output from a 3-input NAND gate would be $Q = \overline{ABC}$ or $Q = \overline{A.B.C.}$

**Note**: it is most important that we add one bar over the whole expression.

## By symbols and a truth table

The circuit diagram and truth table are shown in Figure 5.12.

**Figure 5.12**
The 3-input NAND gate symbol and truth table

Symbol

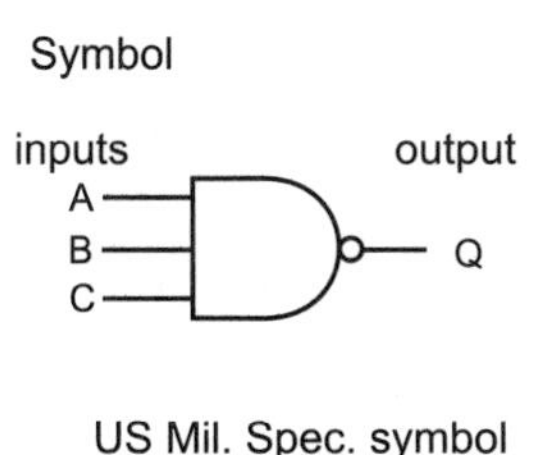


US Mil. Spec. symbol

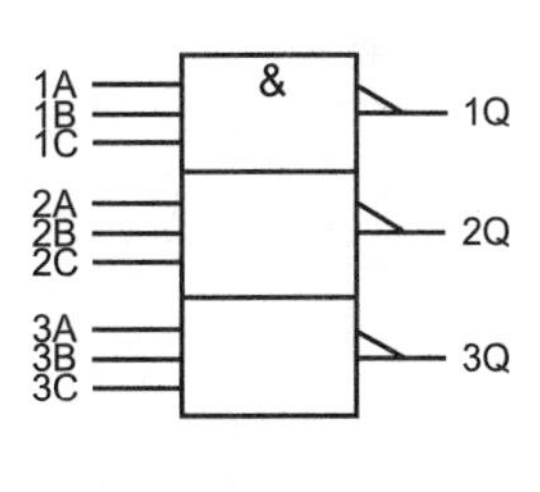


IEC symbol

Truth table

| inputs | | | output |
|---|---|---|---|
| A | B | C | Q |
| 0 | 0 | 0 | 1 |
| 0 | 0 | 1 | 1 |
| 0 | 1 | 0 | 1 |
| 0 | 1 | 1 | 1 |
| 1 | 0 | 0 | 1 |
| 1 | 0 | 1 | 1 |
| 1 | 1 | 0 | 1 |
| 1 | 1 | 1 | 0 |

The output is just the opposite of the AND gate

**An interesting fact**

The NAND gate is used more than any other gate.

## The NOR gate

As we would expect, this is just like an OR gate except for the output being inverted.

## In words

The output is a logic 0 unless all of the inputs are at logic 0.

## As a Boolean expression

Just add a bar across the whole of the normal OR expression. A 2-input NOR gate could be written as $Q = \overline{A + B}$. Remember, once again, that the line extends all the way across the input expression.

## By symbols and a truth table

Just note the inverting circle in Figure 5.13.

**Figure 5.13**
A 2-input NOR gate

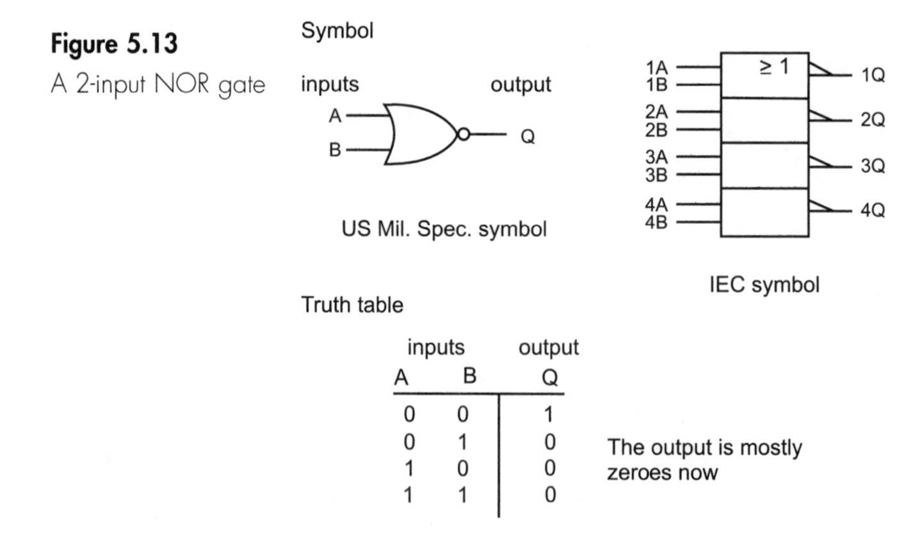


| inputs | | output |
|---|---|---|
| A | B | Q |
| 0 | 0 | 1 |
| 0 | 1 | 0 |
| 1 | 0 | 0 |
| 1 | 1 | 0 |

## The XNOR gate

This is just an inverted XOR gate. This one only has a logic 1 output if the two inputs are the same and, for this reason, it is sometimes called an 'equivalence' gate.

## In words

The output is a logic 1 if the two inputs have the same logic level. Once again, like the XOR gate, only 2-input versions are available.

## As a Boolean expression

Just add a bar across the whole of the normal XOR expression. The output could be written as $Q = \overline{A \oplus B}$.

## By symbols and a truth table

These are shown in Figure 5.14.

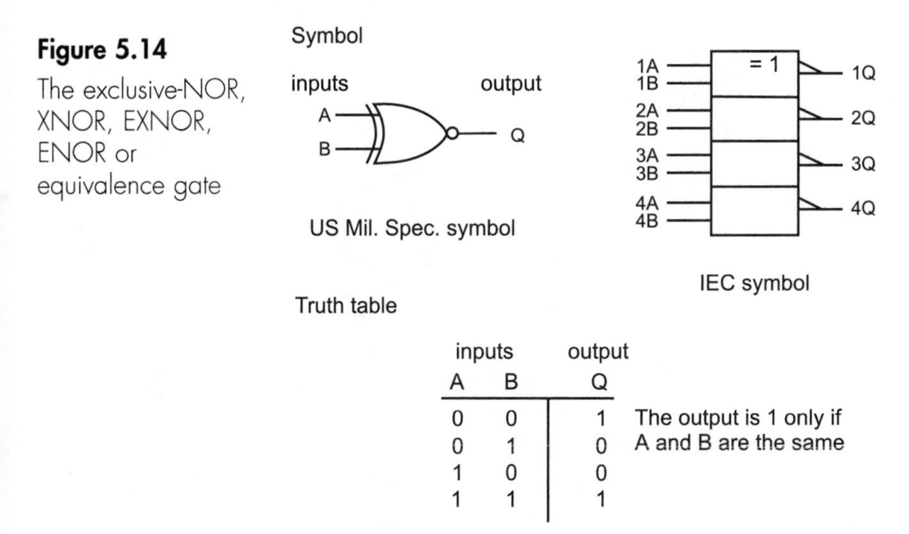


| inputs A | inputs B | output Q |
|---|---|---|
| 0 | 0 | 1 |
| 0 | 1 | 0 |
| 1 | 0 | 0 |
| 1 | 1 | 1 |

**Figure 5.14**
The exclusive-NOR, XNOR, EXNOR, ENOR or equivalence gate

## Real gates

For less than the price of a can of Coke®, you can treat yourself to a NAND gate.

What you will get for your money will be like the creature shown in Figure 5.15. This one has 14 pins, which are used to make the electrical connections to the remainder of the circuit. The device would plug into a socket called a 'base', or it could be soldered directly onto the printed circuit board.

The logic gates are in integrated form. That means that all the electronic components, like transistors, are built within a solid piece of silicon called an integrated circuit (IC or chip). It is only a couple of millimetres in size and is too small to handle, so several gates are included in a single plastic moulding called a 'package'.

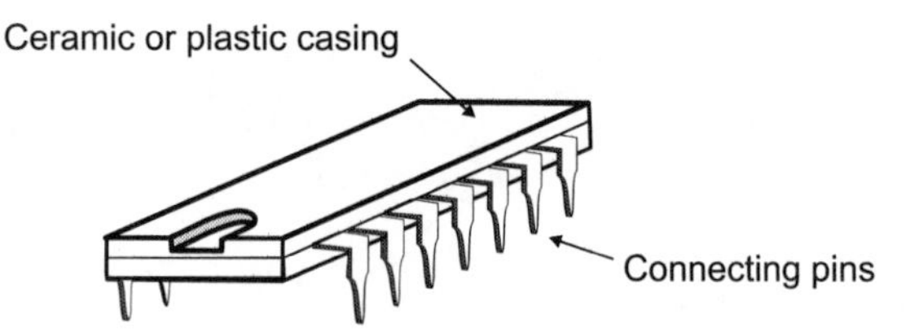


**Figure 5.15**
A typical digital package

A two-input NAND gate, for example, would require two of the pins to connect the input voltages and one to provide the output. That's three altogether. Four such gates would need 12 pins. The 'chip' or 'package' has 14 pins. The two extra ones are used for the power supply. The connections are shown in data books as in Figure 5.16.

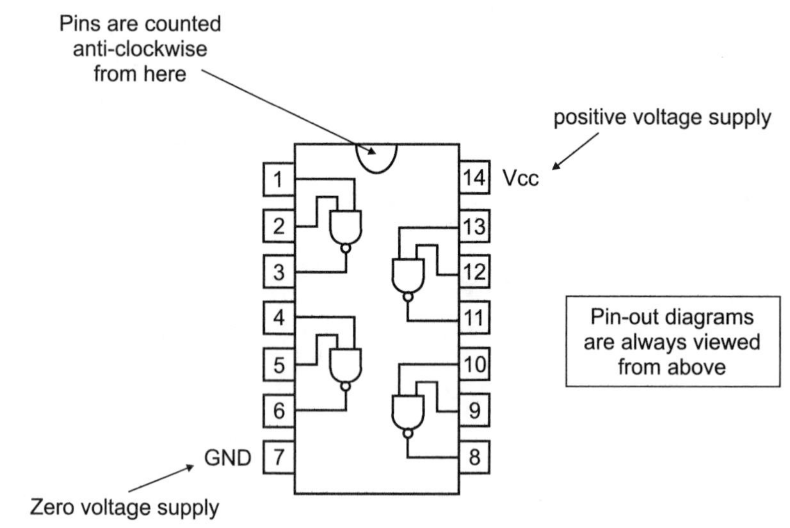


**Figure 5.16**
The pins on a 2-input NAND gate

If we wanted a 3-input gate, then each gate would occupy three inputs and the one output, a total of four pins. The package would then hold three such gates, plus the power supplies.

A 13-input NAND gate cannot get into a 14-pin package, however hard we push. Standard base sizes for the smaller logic circuits are 14, 16, 18, 20, 24 and 28.

Sometimes the arithmetic is not so convenient as in our examples above and we finish up with some unused pins – these are just left unconnected, and on pin-outs are labelled as n.c. (not connected).

We are supplied with the so-called 'pin-out' diagrams by the manufacturers to let us know the purpose of each pin. Sometimes they include small gate symbols as in Figure 5.16, or letter symbols can be used. The labelling is not standardized between manufacturers.

The pins are numbered looking from the top of the IC starting from the left-hand corner nearest to the indentation at the end of the moulding. The pins are then counted anticlockwise around the IC, as we saw in Figure 5.16. Be careful not to mistake the indentation with a circular mark that is sometimes left from the moulding process. Notice also that the pin layout is symmetrical, which means that the integrated circuit can be plugged in the wrong way round – which is immediately fatal to the integrated circuit.

You will have noticed that the power supplies were connected to pins 7 and 14 so the bottom left-hand pin is connected to zero volts and the top right-hand pin is the positive supply. On other packages the same relative positions are normally followed, so, for example, on a 16-pin package we would find the zero volt supply connected to pin 8 and the positive supply on pin 16. The use of these pins is fairly standard but is NOT universal – so be careful. Incorrect power connections kill the chip!

## Quiz time 5

In each case, choose the best option.

**1 A gate with an output of Q = A + B is:**

(a) an AND gate.
(b) a NOR gate.
(c) a 2-input XOR gate.
(d) a 2-input OR gate.

**2 The symbol $\oplus$ indicates:**

(a) an XOR gate.
(b) an AND gate.
(c) a NOT gate.
(d) a NOR gate.

**3 Which of these groups refer to the same gate?**

(a) EOR, EXNOR, NOR.
(b) EOR, XOR, EXNOR, difference gate.
(c) ENOR, EXNOR, XNOR, equivalence gate.
(d) XNOR, ENOR, difference gate, EXNOR

**4 How many NOT gates would a 14-pin package contain?**

(a) 6
(b) 7
(c) 12
(d) 14

**5 A 2-input gate has inputs of A = 1 and B = 0; if the output is a 1, the gate could be:**

(a) an AND, ENOR or a NOR gate.
(b) an AND, XOR or an NOR gate.
(c) a NAND, XOR or an OR gate.
(d) a NAND, XNOR or an OR gate.

# 6

# Build your own gates

Along with the law of nature that decrees that buttered toast always lands butter-side down, there is one that states 'However many logic gates we have, the one we want is not amongst them'.

Unlike the toast, we usually have an easy solution to the logic problem.

How would you add the numbers $3+4+5$?

You could say, but it is unlikely, that you just added all three at once. It is more likely that you added the 3 and the 4 to give 7, and then added 7 and 5 to give the required total of 12. Note that we could have got to the same answer by adding any two and then the third number afterwards. This 'any order' feature is called the 'commutative' property.

The same occurs with logic gates. If we wanted to perform the logic function A OR B OR C, written as $A+B+C$, we can do $A+B$ first, then combine the result with C afterwards.

We could put brackets around the $(A+B)$ to show that this was done separately and requires a 2-input OR gate, and combining the result with the C input would need another 2-input OR gate. The method is shown in Figure 6.1.

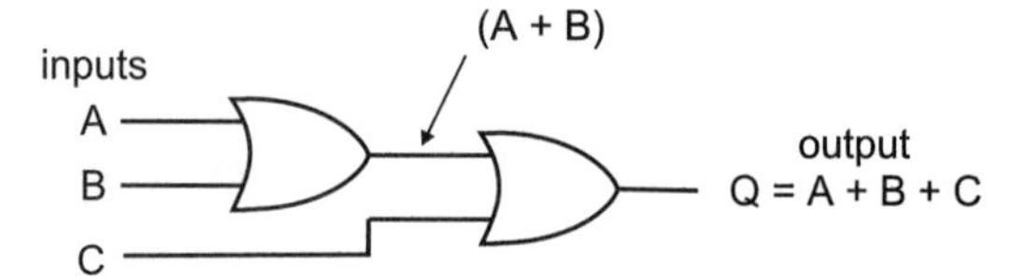


**Figure 6.1**
A do-it-yourself 3-input OR gate

**Example**

Show how you could produce a 4-input AND gate from two 3-input AND gates.

Answer:

The method is much the same as we used with the OR gates. In this case we can AND three of the inputs and then AND the other one. This would be written as (ABC)D or, if you prefer, (A.B.C).D.

Have a look at Figure 6.2.

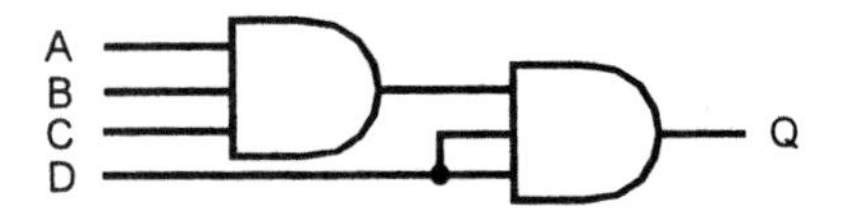


**Figure 6.2**
Building a 4-input AND gate

## Reducing the number of inputs

How could we use a 3-input OR gate as a 2-input gate? A 3-input OR gate has a truth table as shown in Figure 6.3. We can see that, if any

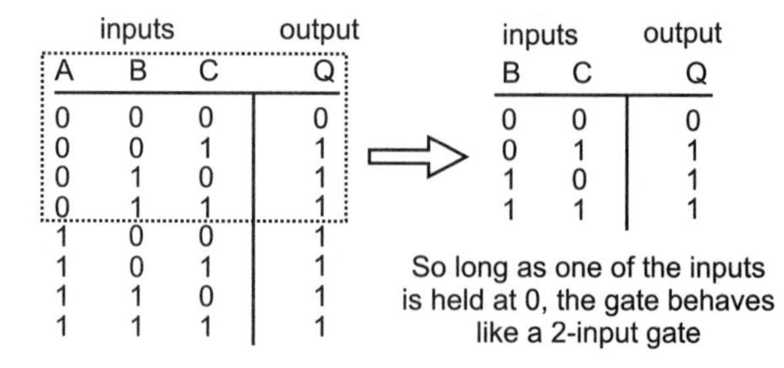

| inputs | | | output |
|---|---|---|---|
| A | B | C | Q |
| 0 | 0 | 0 | 0 |
| 0 | 0 | 1 | 1 |
| 0 | 1 | 0 | 1 |
| 0 | 1 | 1 | 1 |
| 1 | 0 | 0 | 1 |
| 1 | 0 | 1 | 1 |
| 1 | 1 | 0 | 1 |
| 1 | 1 | 1 | 1 |

| inputs | | output |
|---|---|---|
| B | C | Q |
| 0 | 0 | 0 |
| 0 | 1 | 1 |
| 1 | 0 | 1 |
| 1 | 1 | 1 |



**Figure 6.3**
The 3-input to 2-input OR truth table

one of the inputs is held permanently at logic 0, the gate behaves just the same as a 2-input gate. We can achieve a similar result by joining any two of the inputs together, ensuring that two of the inputs are always at the same logic level. See Figure 6.4.

**Figure 6.4**
Two ways of changing a 3-input to a 2-input

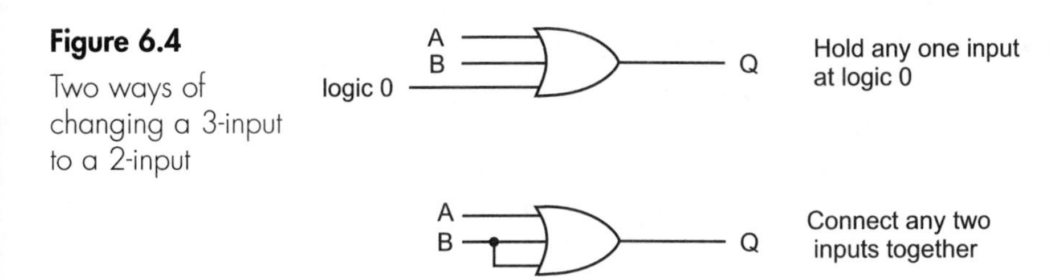


**Example**

Show how you could produce a 2-input AND gate from a 3-input AND gate.

Answer:

**Figure 6.5**
From 3-input to 2-input AND gates

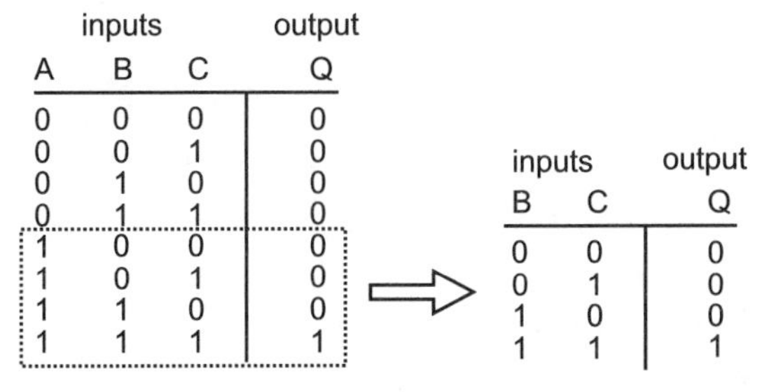

| inputs | | | output |
|---|---|---|---|
| A | B | C | Q |
| 0 | 0 | 0 | 0 |
| 0 | 0 | 1 | 0 |
| 0 | 1 | 0 | 0 |
| 0 | 1 | 1 | 0 |
| 1 | 0 | 0 | 0 |
| 1 | 0 | 1 | 0 |
| 1 | 1 | 0 | 0 |
| 1 | 1 | 1 | 1 |

| inputs | | output |
|---|---|---|
| B | C | Q |
| 0 | 0 | 0 |
| 0 | 1 | 0 |
| 1 | 0 | 0 |
| 1 | 1 | 1 |

This time, one of the inputs must be held at a logic 1 level

The method is much the same as we used with the OR gates – but not quite. We can certainly connect any two of the inputs together but, looking at the truth table in Figure 6.5, it is obvious that we cannot simply connect one of the inputs to logic 0. Once again, we have two methods of achieving the required result. We can either hold one of the inputs at logic 1, or we can simply connect two of the inputs together as shown in Figure 6.6. Once again, it doesn't matter which inputs are used, they all behave in the same way.

**Figure 6.6**
The 3-input AND changes to a 2-input AND

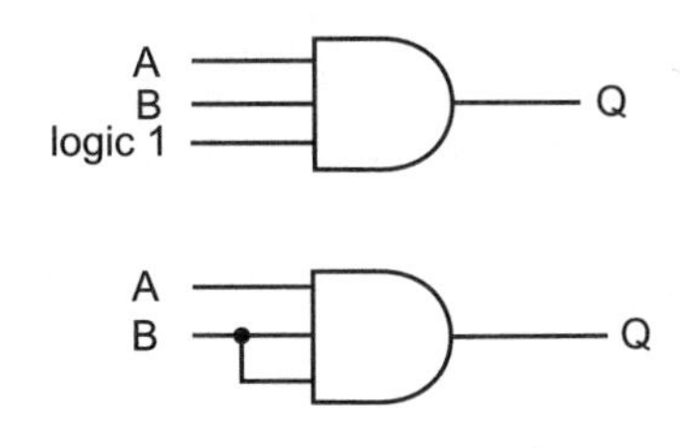


A 3-input AND gate

## Conversions using the NOT gate

The only difference between a NAND gate and an AND gate is the inverter that has been added to the output. We have seen in Chapter 5 that we can cancel the effect of an inverter by adding another one. So, adding an inverter on the end of a NAND gate will convert it back again to an AND gate.

**Example**

How would you convert a NOR gate to an OR gate?

Answer:

Just add a NOT gate as in Figure 6.7.

**Figure 6.7**
NOR to OR, just add a NOT

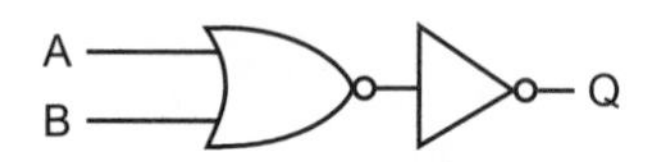


The NOT gate can also change NAND to AND

Here's a more interesting one.

What single gate could be used to replace the combination shown in Figure 6.8?

**Figure 6.8**
What gate is this?

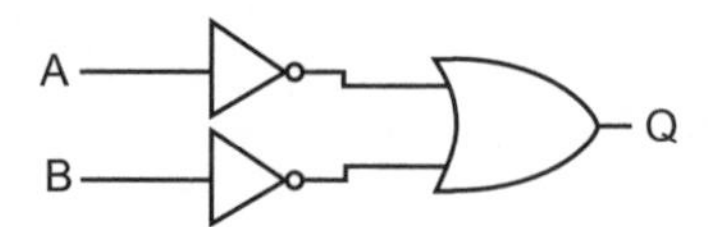


NOT gates can also be added to the inputs of a gate

The effect of adding the NOT gates to the inputs is to invert the inputs before they are applied to the OR gate. This has a surprising effect. The easiest way to see what happens is to make the changes to the truth table, then see if we can recognize it. This is shown in Figure 6.9.

## Other possibilities

If we were to add NOT gates to the input of an AND gate we could draw up the truth tables and discover that the overall result would be a NOR gate.

**Figure 6.9**
A surprising result!

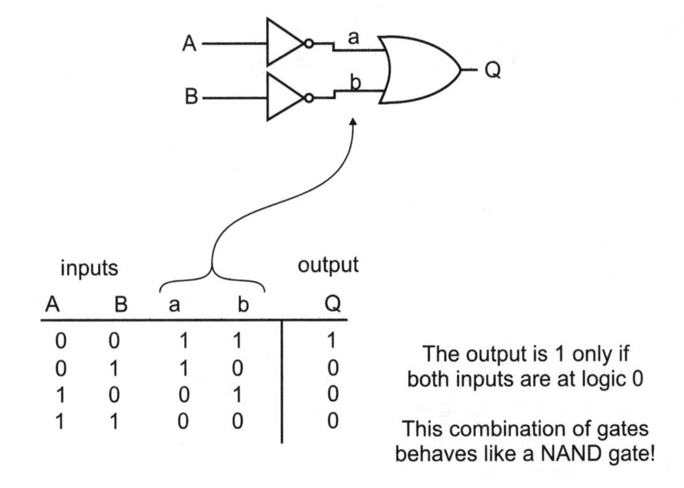


| A | B | a | b | Q |
|---|---|---|---|---|
| 0 | 0 | 1 | 1 | 1 |
| 0 | 1 | 1 | 0 | 0 |
| 1 | 0 | 0 | 1 | 0 |
| 1 | 1 | 0 | 0 | 0 |

Ploughing through all the possibilities would quickly become tiresome, so everything we need to know is in Figure 6.10.

**Figure 6.10**
A very useful thing to know

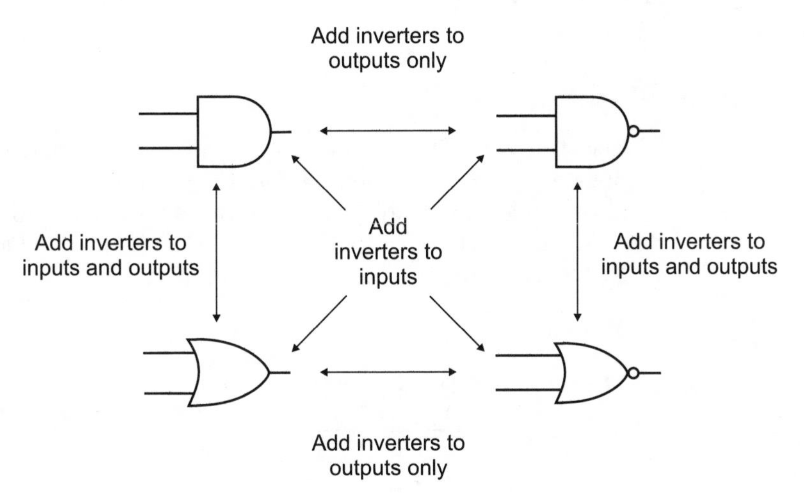


## How to use Figure 6.10

Start from the gate that you have, then move to the gate that you want. Add inverters as written against the arrows. For example, assuming we have a NOR gate, we would start at the bottom right-hand corner. If we wanted to convert it to an AND gate, we would move diagonally across the diagram passing the instruction 'add inverters to inputs'. If, however, we wanted to convert the NOR gate to a NAND gate, we would move vertically and see the instruction 'add inverters to inputs and outputs'.

**Example**

Use the diagram to determine the type of gate represented in Figure 6.11.

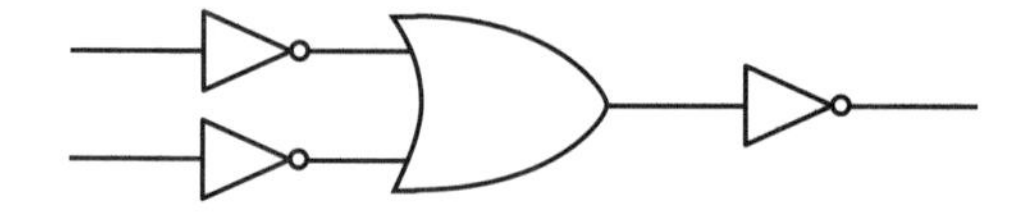

**Figure 6.11**
What gate is this?

Start with the OR gate in the bottom left-hand corner. What has happened to the OR gate? It has had inverters added to the inputs and to the output. So on Figure 6.10, you would move vertically and the result would be an AND gate.

What if you thought it was a NOR gate with inverters added to the input? OK, so you would start at the bottom right-hand corner and then you would have to move along the diagonal to add the inverters at the input. This would give the result as an AND gate just the same. This diagram always works. It's a useful thing to learn.

## Universal gates

If we take a NAND gate and connect all the inputs together, what type of gate do we have? The truth table in Figure 6.12 would be reduced to only two possibilities. A zero level input would provide a logic 1 output, and likewise a logic 1 input would be inverted to a logic 0. A NAND gate can be easily changed to a NOT gate.

**Figure 6.12**
NAND to NOT – instantly

| inputs | | output |
|---|---|---|
| A | B | Q |
| 0 | 0 | 0 |
| 0 | 1 | 0 |
| 1 | 0 | 0 |
| 1 | 1 | 1 |

| inputs | | output |
|---|---|---|
| A | B | Q |
| 0 | 0 | 0 |
| 1 | 1 | 1 |

The truth table is now the same as an inverter

The number of inputs doesn't matter, and a NOR gate would work just as well.

We saw, a moment ago, that we could convert a NAND gate (or a NOR gate) to any other type of gate simply by adding inverters at the output or inputs or both.

If we have a supply of NAND or NOR gates, we can make any other gates. For this reason, the NAND and NOR gates are referred to as Universal gates. As an example, Figure 6.13 shows how to make a NOR gate from NAND gates.

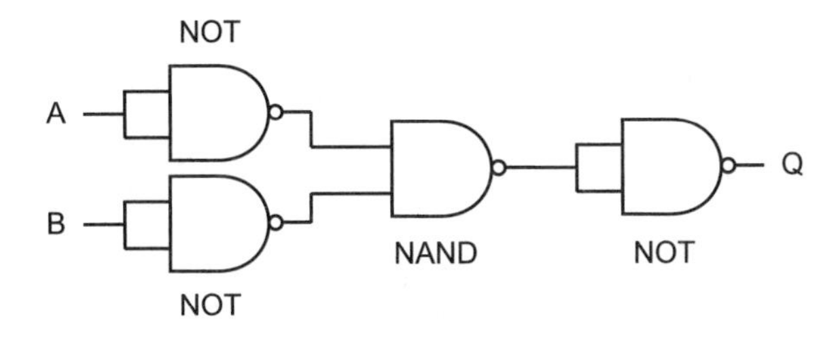


**Figure 6.13**
A NOR gate built from NAND gates

**Example**

Try this one. Show how we could build a 3-input AND gate from a supply of 3-input NOR gates, then check your answer by seeing Figure 6.14.

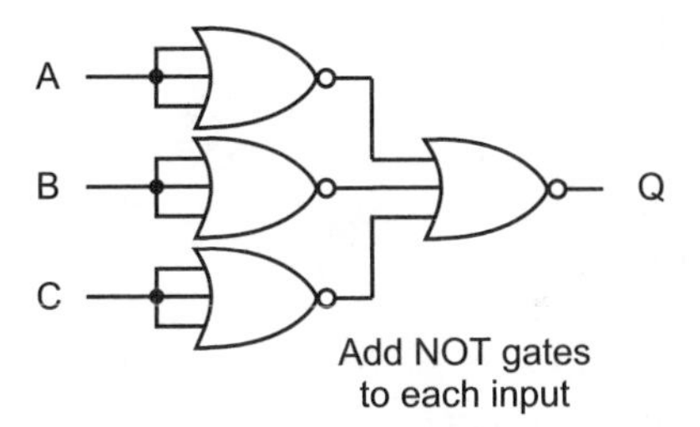


**Figure 6.14**
A 3-input AND from 3-input NORs

## De Morgan's laws

Despite the impressive name, these laws are just the Boolean equivalent of the conversions shown in Figure 6.10. They are used to change expressions using AND and NAND gates into equivalent circuits using OR or NOR gates.

We can see from the figure that if we want a 2-input OR gate we can make it out of a 2 input AND gate by using an AND gate, inverting each input signal and then inverting the final output.

Notice the three steps – change the gate type, invert the inputs, invert the output and that is all there is to it.

De Morgan's law follows exactly the same steps, but we write it in Boolean form.

See this:

Step 1: Change the gate type from OR to AND, so a 2-input OR gate $A+B$ becomes $A.B$

Step 2: Invert each term so $A.B$ becomes $\overline{A}.\overline{B}$

Step 3: Invert the whole expression to give $\overline{\overline{A}.\overline{B}}$

So $A+B = \overline{\overline{A}.\overline{B}}$, which is just what we would achieve by using the conversion in Figure 6.10.

**Example**

We have a 2-input NOR gate. Find the NAND equivalent.

Step 1: Change the symbol: $\overline{A + B}$ to $\overline{A.B}$

Step 2: NOT each term to give $\overline{\overline{A}.\overline{B}}$

**Note**: We could have written the result of this step as $\overline{\overline{A}}.\overline{\overline{B}}$, but it is more usual to stack the bars with the shortest ones at the lowest level.

Step 3: NOT everything to give $\overline{\overline{\overline{A}.\overline{B}}}$

Step 4: Simplify bars of equal length if required.

If we decided to cancel the two equal length bars, the result would be $\overline{A + B} = \overline{A}.\overline{B}$

Now this is certainly correct in as much as it performs the correct logic function. However, we were asked to produce this function using NAND gates but $\overline{A}.\overline{B}$ is using an AND gate. If we replace this AND gate with a NAND gate, the result would be written as $\overline{\overline{A}.\overline{B}}$ but we still have a problem because we have now inverted what was the correct expression so we must invert it back again, which will be shown by another full length bar being added to make the final answer $\overline{\overline{\overline{A}.\overline{B}}}$

The final circuit is shown in Figure 6.15.

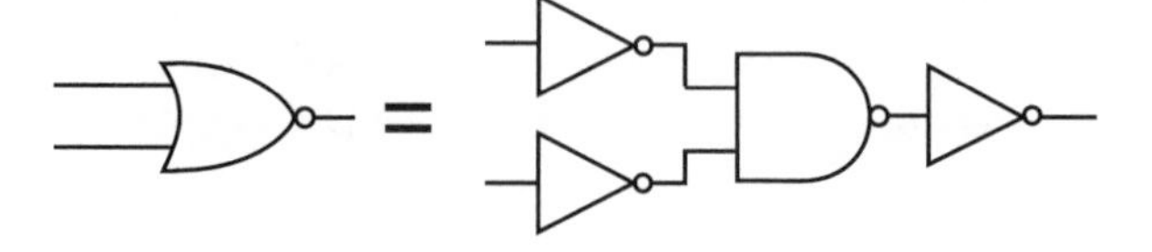

**Figure 6.15**
Making our own NOR gate

**Example**

Change the expression $F\overline{D} + A + \overline{B + C}$ to remove any OR gates.

Step 1: Change the symbols to give $F\overline{D}.A.\overline{B + C}$

Step 2: NOT each one: $\overline{\overline{F\overline{D}}.\overline{A}.\overline{\overline{B}.\overline{C}}}$

Note that $F\overline{D}$ is treated as a single signal. In fact it IS a single signal – it is the output of a 2-input AND gate.

Step 3: NOT the lot! $\overline{\overline{F\overline{D}}.\overline{A}.\overline{\overline{B}.\overline{C}}}$

Step 4: There are no equal length bars stacked one above the other so no simplifications are possible.

## Summary of De Morgan's law

1 Change the symbols.
2 NOT each one.
3 NOT the lot.

### Have the XOR and XNOR gates been forgotten?

No, just ignored, but we can certainly build them from either of the universal gates. Working from the truth tables, XOR is $A\overline{B} + \overline{A}B$ and XNOR is just the same but inverted, $\overline{A\overline{B} + \overline{A}B}$

## Advantages of using universal gates

There are two advantages:

1 We can use up spare gates in the chips that we have already bought instead of buying new ones.
2 If we decide to use just one type of gate, either a NAND or NOR, we can buy them more cheaply by ordering them in large quantities. Most manufacturers have opted for the NAND gate, and hence more NAND gates are bought than any other.

## Quiz time 6

In each case, choose the best option.

**1 How many 'universal' gates are there?**

(a) 6
(b) 4
(c) 3
(d) 2

**2 An AND gate with inverters connected to each input behaves like:**

(a) a NAND gate.
(b) a NOR gate.
(c) an OR gate.
(d) an AND gate.

**3 Using only two 3-input NOR gates, we could NOT build a:**

(a) 3-input OR gate.
(b) 3-input AND gate.
(c) 4-input NOR gate.
(d) 3-input NOT gate.

**4 If we had a supply of AND gates and plenty of NOT gates, we could construct:**

(a) only AND gates.
(b) only NAND gates.
(c) any other gate.
(d) only OR gates or NOR gates.

**5 To change a 3-input NAND gate into a 3-input NOR gate would require:**

(a) only one inverter.
(b) two inverters.
(c) three inverters.
(d) four inverters.

# 7

# Designing digital circuits

Sometimes we buy a digital circuit already designed, built and included in products like CD players, mobile telephones or computers. There is no reason, however, to prevent us from designing and building our own digital circuit to do something that we want.

## Digital circuits are cheap and easy to build

Photographers often need red light or complete darkness to develop a film. They tend to get rather irate if someone opens the door just to see how they are getting on.

A solution to this problem is to install a lock on the door that is controlled by switches that sense whether the red 'developing' light is ON and also whether the normal white room lights are ON, and only allow the door to be opened under safe conditions.

At this moment our 'digitally aware' photographer may decide that there are no instant solutions available in the shops, so a 'home-made' solution is required.

## How do we design a digital circuit?

Easy, in just five steps.

1 Write down what we want the circuit to do.
2 Choose the switches that we want to use.

3 Draw up a truth table.
4 Simplify it if we can; if we can't, just go to step 5.
5 Draw up the logic diagram and then build it.

We should be careful to work slowly, building up the design one step at a time and making notes as we go.

Let's have a go at the photographer's problem using the steps listed above.

Step 1: We would like the door lock to be activated (locked) whenever the red light is ON at the same time as the white light is OFF.

Step 2: Simple on-off switches come in two flavours referred to as 'normally-open' and 'normally-closed'.

A normally-open switch has contacts that are initially separated so that current cannot flow. A normally-closed switch operates the other way round and has the contacts touching allowing current to flow, as illustrated in Figure 7.1. The method of closing the contacts differs according to the design of the switch. Switches can be light sensing, respond to mechanical movement or pressure, or to any one of a dozen different options that can be found in electronic component catalogues.

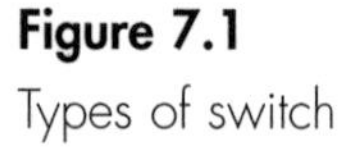


**Figure 7.1**
Types of switch

We will use a normally-open switch for monitoring the red light and a normally-closed switch to monitor the state of the white light. There is no magic about these choices, the circuit could be designed equally well whatever the type of switch used, but one of each will help us to explore the possibilities.

Step 3: The truth tables that we have met so far have used A, B, C etc. for the inputs with Q for the output. It is not desperately important to use these letters, so when we come to our own designs we may find it easier to use letters that will help us to remember what each column of the truth table is all about. So, we can call the switch associated with the red light, R. The white light switch can be W, and the lock can be L.

With just two switches there can only be four possible combinations, as shown below:

| R | W | L |
|---|---|---|
| 0 | 0 | |
| 0 | 1 | |
| 1 | 0 | |
| 1 | 1 | |

where

R = red light ON = 1
W = white light ON = 1
L = room locked = 1.

Now we fill in the last column.

Go back to our original design statement: 'We would like the door lock to be activated (locked) only when the red light is ON at the same time as the white light is OFF.' This would happen on the row R = 1, W = 0, and we complete the line by putting L = 1 to show the door is locked. In all other cases the door is unlocked, so cups of coffee and sandwiches can be brought in at any time. These conditions are listed as L = 0.

| R | W | L |
|---|---|---|
| 0 | 0 | 0 |
| 0 | 1 | 0 |
| 1 | 0 | 1 |
| 1 | 1 | 0 |

Step 4: We will leave the problem of simplification to the next chapter.

Step 5: Looking at the final version of the truth table, we simply ignore any lines that have an output of '0'. So that has disposed of three out of the four lines.

This has shrunk the table to:

| R | W | L |
|---|---|---|
| 1 | 0 | 1 |

## How do we convert a truth table into a logic diagram?

The line in the truth table means that the L (lock) is activated when the R (red light) is on AND the W (white light) is off. This indicates an AND gate is to be used. In fact, any combination of gates in a single row is always an AND function. Our particular example has two columns, and so it is a 2-input AND gate.

So we start by drawing an AND gate as in Figure 7.2.

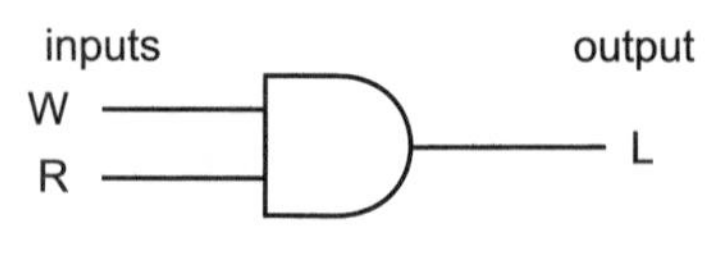


**Figure 7.2**
Our first attempt

This diagram would not suit our purposes. An AND gate requires both inputs to be at a logic 1, and so the lock would only be activated by both the white light and the red light being ON.

Even so, we need an AND gate because there are two things that must occur at the same time, but it must operate when the white light is OFF.

### We have a problem

When the white light is OFF, its switch has an output of logic 0 but, to make it work, the AND gate needs a logic 1 – so what do we do?

The answer is to include a NOT gate to invert the logic level produced by the white light switch. That gives the logic 1 needed for the AND gate and produces a final diagram as shown in Figure 7.3.

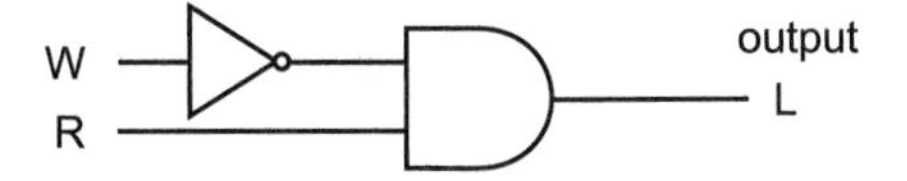


**Figure 7.3**
The final solution

The lock is now activated when the white light is OFF and the red light is ON

**Example**

A different situation results in this truth table:

| A | B | C | Q |
|---|---|---|---|
| 0 | 0 | 0 | 0 |
| 0 | 0 | 1 | 1 |
| 0 | 1 | 0 | 0 |
| 0 | 1 | 1 | 0 |
| 1 | 0 | 0 | 0 |
| 1 | 0 | 1 | 1 |
| 1 | 1 | 0 | 0 |
| 1 | 1 | 1 | 1 |

## Design a logic diagram to provide this result

Step 1: There are three rows that result in $Q = 1$. To put it another way, we can say that $Q = 1$ under the conditions described in the first of these rows OR in the second OR in the third. These conditions are independent of each other, and this means that the final gate will be a 3-input OR gate.

Step 2: Cross out or ignore all the rows that result in the output Q = 0; all these are ignored.

| A | B | C | Q |
|---|---|---|---|
| 0 | 0 | 1 | 1 |
| 1 | 0 | 1 | 1 |
| 1 | 1 | 1 | 1 |

Our logic diagram now looks like Figure 7.4.

**Figure 7.4**
The first step

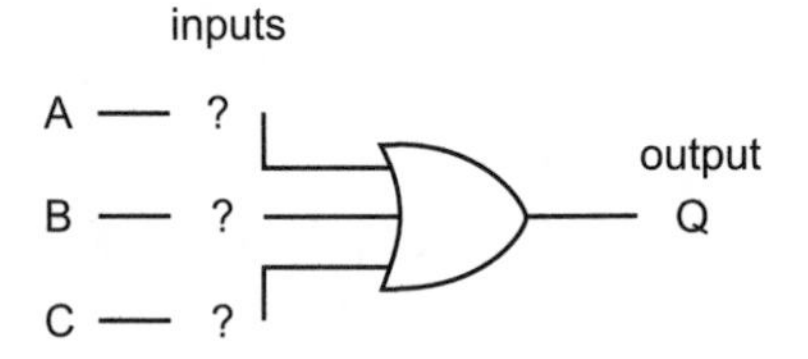


So far, we know that there are three inputs and the last gate is a 3-input OR gate

Step 3: The first row has A = 0 AND B = 0 AND C = 1. This means that it is a 3-input AND gate. Now, to operate this AND gate we need the three inputs to be at logic 1. However, A = 0, so we must add an inverter to produce the logic 1 level. Input B must also be inverted for the same reason. Figure 7.5 shows the first row completed.

**Figure 7.5**
The first row is completed

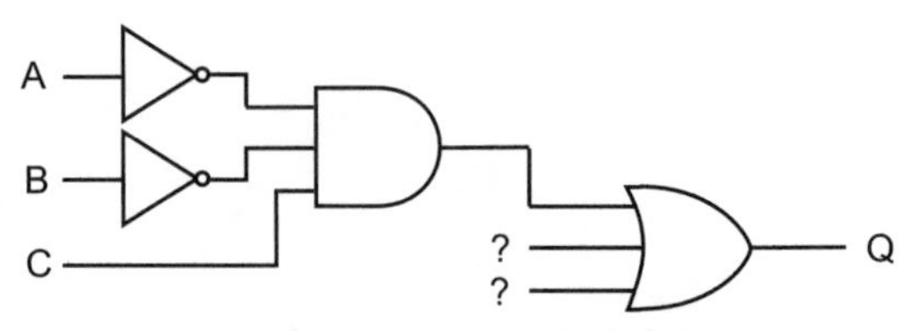


When A = 0 and B= 0 and C = 1, there will be three level 1 inputs to the AND gate

Step 4: The next row is much the same situation except that input A = 1 and C = 1 and so do not need inverting. Column B will still need a NOT gate as before. Figure 7.6 shows this next addition to the logic diagram.

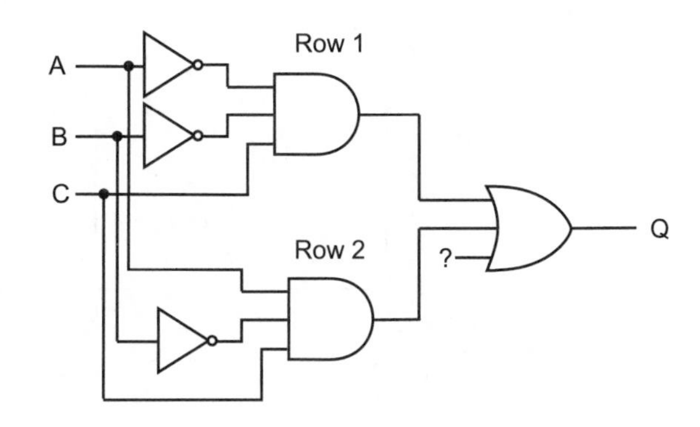


**Figure 7.6**
And now the second row is added

Step 5: The third row is easy. It is a straightforward 3-input AND gate. No inverters are needed because all columns already have a logic value of 1. The final logic diagram is shown in Figure 7.7.

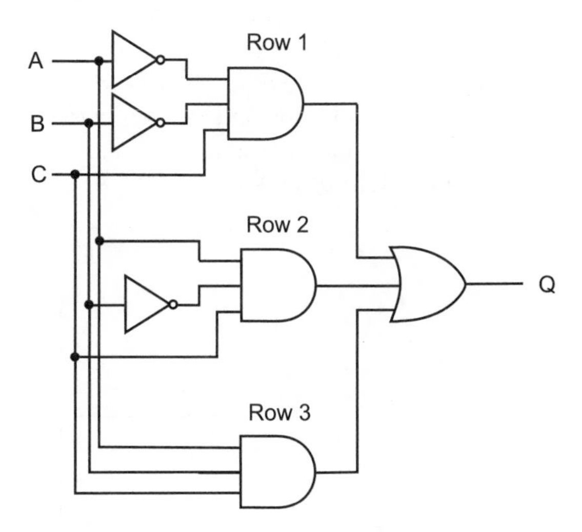


**Figure 7.7**
Complete – and very impressive

**Note**: In Step 4, we added a NOT gate to provide the NOT B input to the Row 2 AND gate, but if we look at the diagram we can see that we already have a NOT B logic level being fed into the Row 1 AND gate. We could save a NOT gate by using this logic level rather than use another NOT gate. In Figure 7.8 the alternative layout is shown.

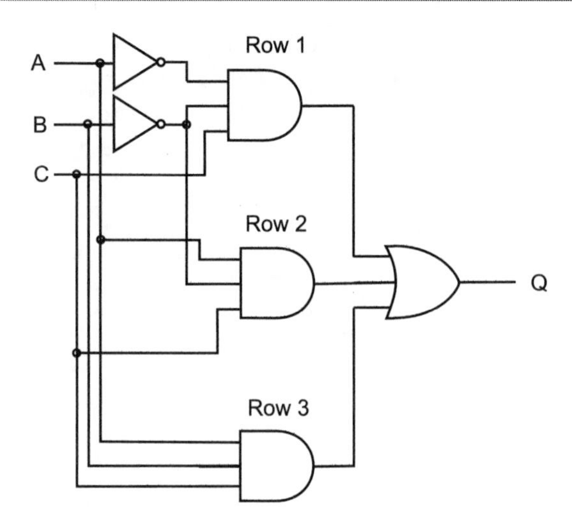


**Figure 7.8**
An alternative design

## These are the main points

1 To convert any truth table to a logic diagram, start by drawing an OR gate at the output. The number of inputs to the OR gate is determined by the number of 1s in the output column. Each input is fed by an AND gate.
2 The number of inputs to each AND gate is equal to the number of input columns in the truth table. Each input that is at a logic 0 must be fed via a NOT gate.
3 Whenever possible draw the logic diagram so that the inputs appear on the left-hand side and the output to the right, so we can read the diagram from left to right. Be sure to make it clear whether lines on diagrams join or cross, as in Figure 7.9.

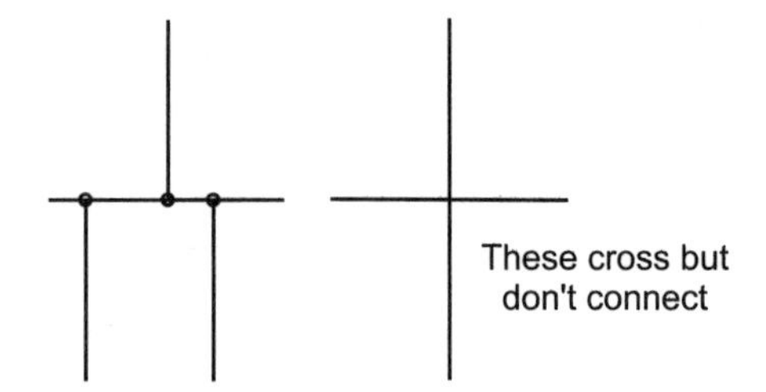


**Figure 7.9**
Junctions and crossings

4 The order in which we write down the separate letters or the different groups of letters does not matter.

**Example**

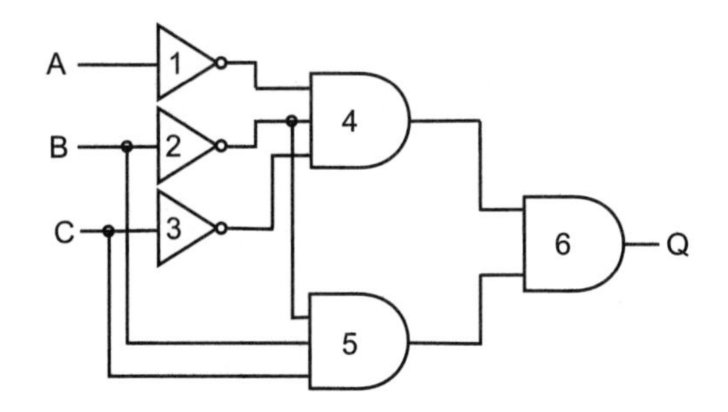


**Figure 7.10**
Spot the mistakes

Here is a truth table. A logic circuit design is shown in Figure 7.10 but it contains mistakes – what are they? The answer is given below.

| A | B | C | Q |
|---|---|---|---|
| 0 | 0 | 0 | 1 |
| 0 | 0 | 1 | 0 |
| 0 | 1 | 0 | 0 |
| 0 | 1 | 1 | 1 |
| 1 | 0 | 0 | 0 |
| 1 | 0 | 1 | 0 |
| 1 | 1 | 0 | 0 |
| 1 | 1 | 1 | 0 |

**Answer:**

The input to gate 5 should be connected to the output of gate 1 and not to the output of gate 2, and gate 6 must be an OR gate.

## How do we convert a truth table into Boolean algebra?

The method is much the same as we used to design the logic diagram.

Here is a truth table:

| A | B | C | Q |
|---|---|---|---|
| 0 | 0 | 0 | 1 |
| 0 | 0 | 1 | 1 |
| 0 | 1 | 0 | 0 |
| 0 | 1 | 1 | 0 |
| 1 | 0 | 0 | 0 |
| 1 | 0 | 1 | 1 |
| 1 | 1 | 0 | 0 |
| 1 | 1 | 1 | 1 |

Step 1: There are four rows that result in Q = 1. These rows are ORed as we saw with the logic diagram.

Step 2: Cross out or forget about all the rows that result in the output Q = 0, as all these are ignored.

Here are the important rows.

| A | B | C | Q |
|---|---|---|---|
| 0 | 0 | 0 | 1 |
| 0 | 0 | 1 | 1 |
| 1 | 0 | 1 | 1 |
| 1 | 1 | 1 | 1 |

Step 3: The first row has $Q = 1$ if $A = 0$ AND $B = 0$ and AND $C = 0$. This means that it is a 3-input AND expression. The Boolean expressions that need to be ANDed are $\overline{A}$, $\overline{B}$ and $\overline{C}$.

The first row is therefore written as $Q = \overline{A}\,\overline{B}\,\overline{C}$.

Step 4: The next row is very similar except that the $C = 1$ and therefore does not need a bar over the C (the 'bar' is the line over the top to indicate the NOT function). The output of this line is $Q = \overline{A}\,\overline{B}C$.

Step 5: The third row is just the same except that it is only the B value that is inverted. The output is now: $Q = A\ \overline{B}\ C$.

Step 6: The last line is a straightforward 3-input AND gate, so the Boolean expression is $Q = A\,B\,C$.

Step 7: All four of these lines need to be ORed together, so the final Boolean expression is $Q = \overline{A}\,\overline{B}\,\overline{C} + \overline{A}\,\overline{B}\,C + A\,\overline{B}\,C + A\,B\,C$.

Brackets are sometimes added if we want to make certain that the terms to be ANDed and those to be ORed are quite clear, so the Boolean could be written as $Q = (\overline{A}\,\overline{B}\,\overline{C}) + (\overline{A}\,\overline{B}\,C) + (A\,\overline{B}\,C) + (A\,B\,C)$.

Step 8: If we can write a Boolean description of a problem, we can go straight into the logic design without worrying about doing the truth table at all. We will have a look at an example in a moment.

## Most important note

Be exceedingly careful to write the bars over each letter separately and not as a single bar over the whole expression. $\overline{A}\ \overline{B}$ is not the same as $\overline{A\,B}$; in fact it is completely different. Try drawing up the truth tables.

## From Boolean to logic

### Example

Draw the logic diagram for the Boolean expression $R = (F\ \overline{K}\ P) + (F\ \overline{K}\ \overline{P})$.

Step 1: The expression contains two groups that are ORed together. We can make a start by drawing a 2-input OR gate. Each of the two inputs is being fed by 3-input AND gates as shown in Figure 7.11.

Step 2: The first AND gate is fed with F, NOT K and P. This means that we must add a NOT gate to the K input to give the value $\overline{K}$. Figure 7.12 shows the first AND gate signal completed.

Step 3: The other AND gate is much the same except that both the K and P inputs are inverted. Figure 7.13 shows the final result.

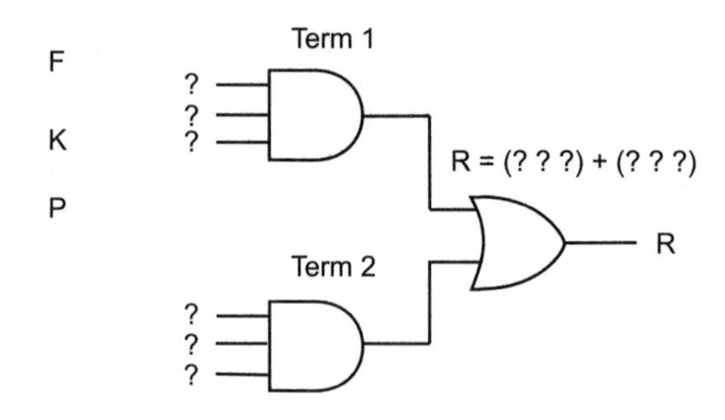


**Figure 7.11**
Boolean to logic – the first step

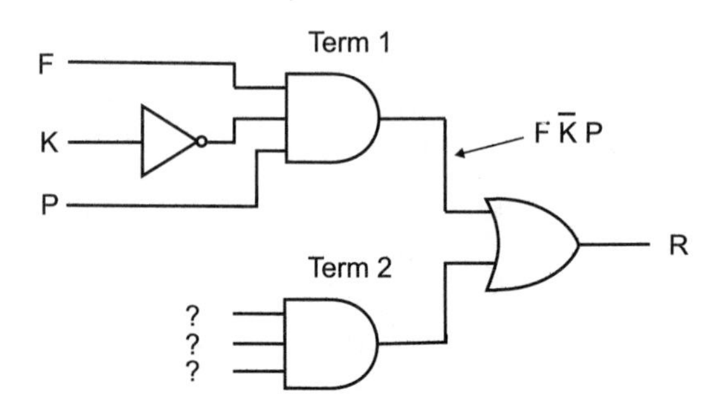


**Figure 7.12**
The first half is finished

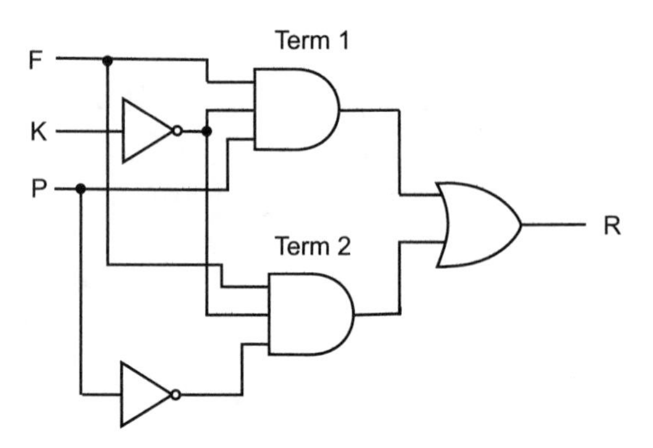


**Figure 7.13**
Boolean to logic – the final result

**Example**

Draw the logic diagram of a circuit that would perform the function $Q = \overline{A}\,\overline{B} + A\,B + A\,\overline{B}$.

Again we have the same basic layout, a 3-input OR gate being fed by AND gates. This time there are three 2-input AND gates.

The first AND gate has an inverter on both the A and the B signal inputs. The second AND gate has no inverters, and the last AND gate has an inverter applied just to the B input.

The final result is shown in Figure 7.14.

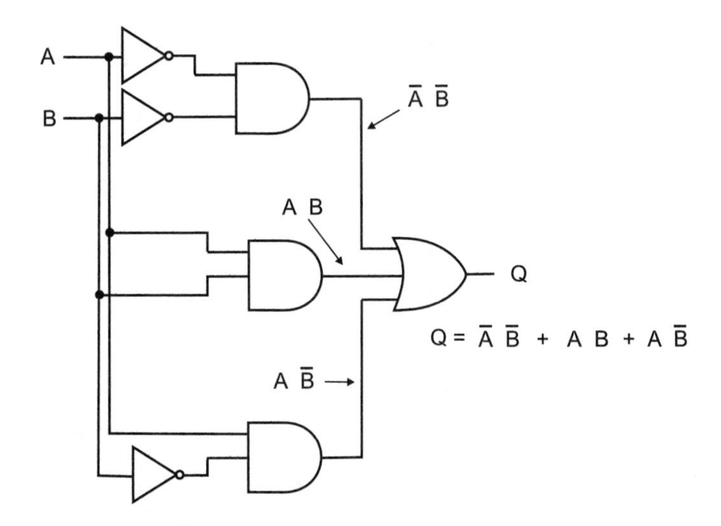


**Figure 7.14**
A worked example

## Other Boolean expressions

In the previous cases we have derived the Boolean expression for truth tables, and these give rise to a standard pattern of AND gates feeding an OR gate. Furthermore, all the AND gates have had the same number of inputs.

This is not always the case. If we start simplifying the Boolean algebra or we build up the expression without starting with a truth table, the final expressions are much more varied. We look at simplification in Chapters 8 and 9.

Let's look at a few other Boolean expressions and see how to draw their logic diagrams. There are, of course, millions of different possibilities, so we will do just a few. To sort these out easily, it is much simpler if we know the Boolean equivalent of the basic gates. If you feel unhappy about them, just glance back at Chapter 5 again.

**Example**

Draw the logic diagram equivalent to the Boolean $Q = A + BC$.

This can be a tricky one. The algebra contains an OR gate and an AND gate, but which do we do first? Is this (A OR B) and the result ANDed with C, or is it A ORed with the result of ANDing B AND C?

To sort out this problem we use the convention that, given the choice, we always do the AND functions first, so the correct interpretation of the above is to AND the B and C and then OR it with the A input.

If we definitely mean it to be done the other way with the A OR B done first then the result ANDed with C, we have to enclose the things to be done first with brackets like this: $Q = (A + B)\,C$.

**Figure 7.15**
Notice the difference

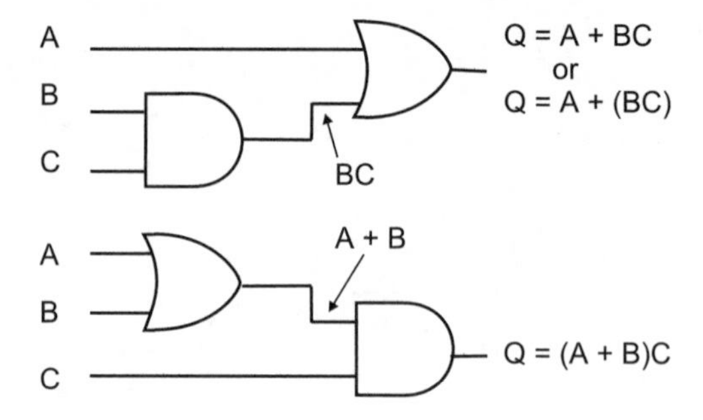


The logic diagram for $Q = A + BC$ is shown in Figure 7.15.

**Example**

Try drawing the logic diagram for the expression $Q = (AB + \overline{AB})C$.

Step 1: It looks complicated. Let's start with a general look at the expression. There is a group of gates inside the brackets and that group has been ANDed with C, so let's draw that first (Figure 7.16).

**Figure 7.16**
The first part

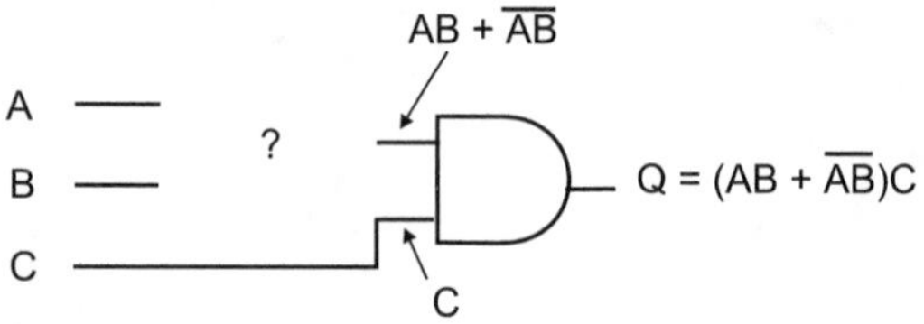

Step 2: Inside the bracket we have AB, which is a 2-input AND gate ORed with $\overline{AB}$, which is a 2-input NAND gate (Figure 7.17).

**Figure 7.17**
The complete diagram

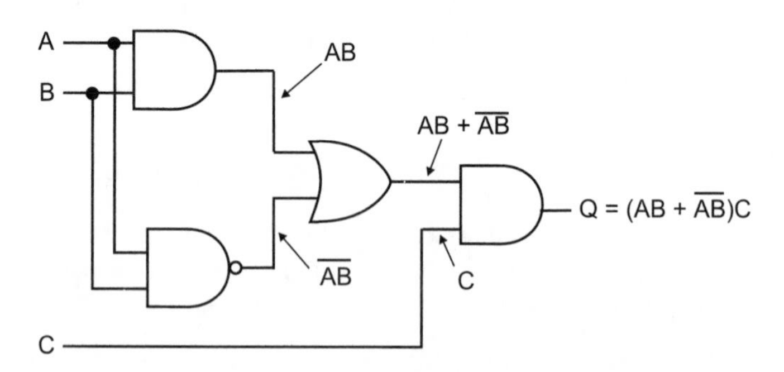


**Example**

Try drawing this one: $Q = \overline{A + BC} + B$.

**Note**: An inverting line over a group like $\overline{A + BC}$ means that the terms A + BC are a group and have then been inverted. This would mean exactly the same as putting it in brackets like this: $Q = (\overline{A + BC})$.

Step 1: The overall pattern is a group of terms ORed with B, so we would have to start with a 2-input OR gate. One of the inputs to this gate is B and the other is the group that has been inverted (Figure 7.18).

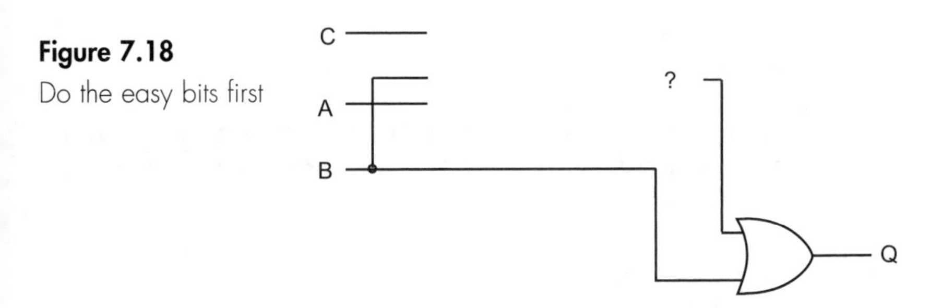


**Figure 7.18**
Do the easy bits first

Step 2: Looking at the inverted group $\overline{A + BC}$, we will build this up in stages. We may spot that this follows the same pattern as a NOR gate.

Remember to do the AND functions first, so B and C go through an AND gate. This takes care of the 'BC' part.

Now we combine this output by using a 2-input NOR gate, using A as the other input. We have now finished A + BC. The complete logic diagram is shown in Figure 7.19.

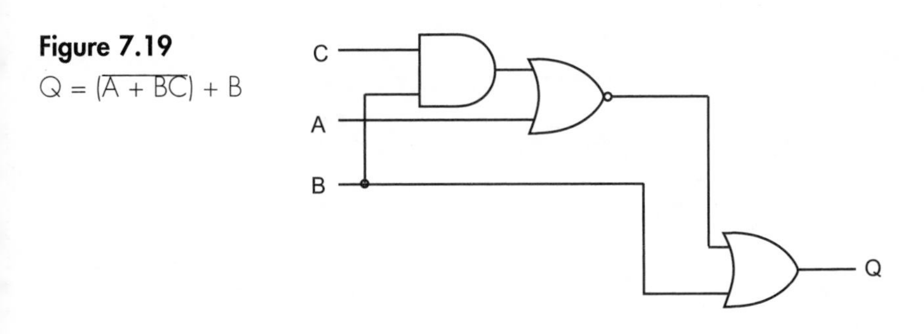


**Figure 7.19**
$Q = (\overline{A + BC}) + B$

**Note**: If you didn't spot that this was a NOR gate, you could have used an OR gate to combine A + BC and then added an inverter to produce the final result $\overline{A + BC}$. It uses an extra gate, which is not quite so slick, but it would certainly work.

## Quiz time 7

In each case, choose the best option.

**1 Which one of the following is the odd one out?**

(a) $C + AB$
(b) $A.B + C$
(c) $(A.B) + C$
(d) $A.(C + B)$

**2 The least number of gates that could be used to build a logic circuit with the Boolean expression $\overline{A + B + C}$ would be:**

(a) 1
(b) 2
(c) 3
(d) 4

**3 The Boolean expression $A + \overline{B}.\overline{C}$ is the same as:**

(a) $A + \overline{BC}$
(b) $\overline{A} + \overline{B}C$
(c) $(\overline{C}\,\overline{B}) + A$
(d) $\overline{B} + C.\overline{A}$

**4 The logic diagram in Figure 7.20 is equivalent to the Boolean expression:**

(a) $H = \overline{F + G.W}$
(b) $H = \overline{F} + \overline{WG}$
(c) $H = \overline{F + W}\,\overline{G}$
(d) $H = (\overline{F + G})W$

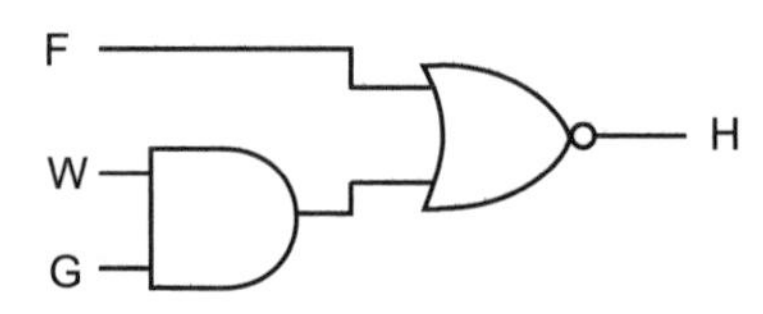


**Figure 7.20**
Quiz time 7, Question 4

**5 In Figure 7.21, the points that will always be at the same voltage are:**

(a) G and E.
(b) A and B.
(c) A and E.
(d) H and A.

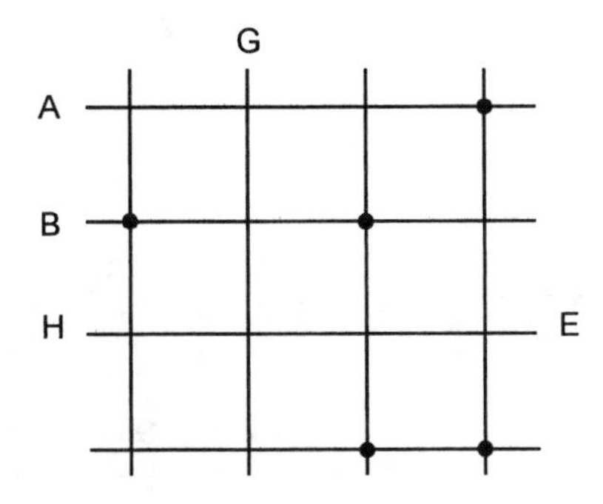


**Figure 7.21**
Quiz time 7, Question 5

# 8

# Simplifying – by Boolean algebra

When we have designed a logic diagram and written it as a Boolean expression, it is usually worthwhile to see if it can be simplified before it is finally constructed.

There are two main methods at our disposal. We can do it by Boolean algebra as in this chapter, or by a drawing method as in the next chapter.

## Why bother to simplify at all?

If we were building a circuit from ready-made integrated circuits, it would typically contain four 2-input NAND gates in each integrated circuit. A circuit requiring 100 NAND gates would need 25 integrated circuits, but if we could simplify the design down to the point where we only need 40 NAND gates we then only need to use 10 chips.

However, there are situations in which we don't bother to simplify at all. In fact, people who follow this route often discard simplification as an outdated process.

In some industrial situations we design and build a single chip that can contain, say, 1000 extremely small NAND gates, which can then be internally interconnected to provide our logic circuit. Our problem is then deciding how to interconnect the NAND gates to provide the required circuit. Let's say the logic circuit requires 800 NAND gates without any simplification, or 600 after simplification. In this situation, there is no point in bothering to simplify the circuit. This would result in 400 unused NAND gates if we simplify and 200 if we don't. But we still use and power up a single integrated circuit and its 1000 tiny NAND gates, so no savings would be made.

## The cost benefits of simplifying

The overall savings may well be significantly greater than the initial cost of the integrated circuits:

1. Integrated circuits. Reducing the number of chips required from 25 to 10 saves the cost of 15 chips.
2. Printed circuit size. The integrated circuits are mounted on a printed circuit board. Reducing the number of chips reduces the size and hence the cost of the PCB. We will also achieve a reduction in the design costs for the board.
3. Sockets. In some pieces of equipment where maintenance is likely to be needed, the integrated circuits are plugged into a socket rather than being soldered directly to the PCB. These sockets often cost as much or more than the integrated circuit.
4. Power supply. If there are fewer integrated circuits, then the power supply design cost, construction cost, size and heat output can all be reduced.
5. Product design. The overall effect is that the size, weight, initial cost and running costs can all be reduced. Less heat and fewer chips would increase reliability.

## Boolean rules OK

Using Boolean algebra may sound impressive and just a bit terrifying, but it is quite a fun thing to do. It allows us to simplify a logic circuit without doing any drawing of the gates. There are only four or five laws used and these are very simple; we quickly get use to using them even though they seem a bit daunting at first.

## Identities

Identities are the most used form of simplification. To understand these, all we need is to feel happy and confident with the truth tables for the basic gates.

There are four identities associated with the OR gate. In each case they show how an OR gate can be removed from the circuit with no ill effects. If you become unhappy, just look back to the truth tables.

### Unnecessary OR gates

If we have a 2-input OR gate with one input called A and the other input held at logic 0, the output will always be the same logic level as the A input. This means that if A = 0 the output is at 0, and if A = 1 then the output is also at 1, and so the identity could be written as A + 0 = A. These two situations are shown in Figure 8.1.

**Figure 8.1**
The identity A + 0 = A

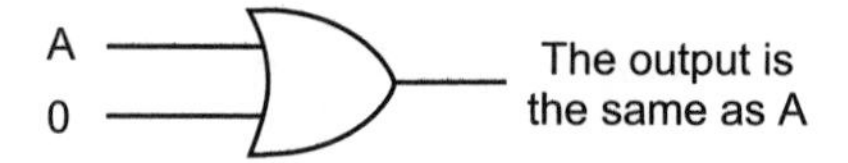

A
0
The output is the same as A
Since the output is the same as the input, this gate is not necessary and the circuit can be simplified to a single piece of wire
A

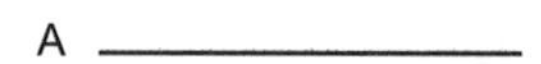

If we build the same circuit but have inputs of A and 1, then the output will always be held at logic 1 whatever the value of A. We now have the simplification A + 1 = 1 as in Figure 8.2. You see how easy the identities are.

**Figure 8.2**
The identity A + 1 = 1

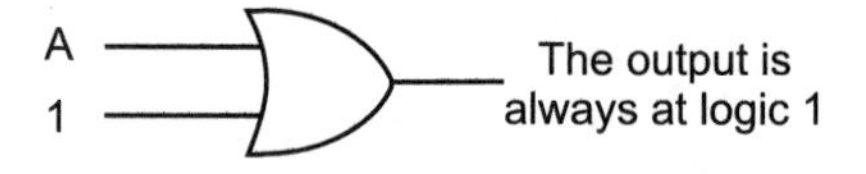

A
1
The output is always at logic 1
So, once again, this gate is unnecessary
1
1

If both inputs are held at level A, then the same logic level is applied to both inputs and so the output will also be at this level. Two 0s give a 0 out, and two 1s give a 1 out. So A + A = A, as in Figure 8.3.

**Figure 8.3**
The identity A + A = A

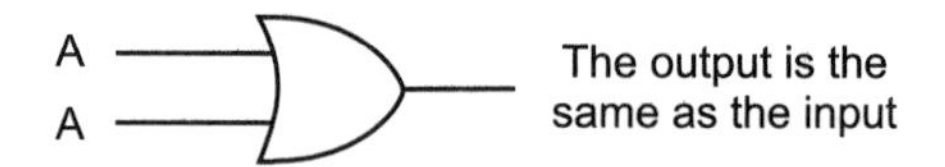

A
A
The output is the same as the input
This gate does nothing and so can be removed
A
A

If one input is at level A and the other is at the opposite value, $\overline{A}$, then one of them will always be at a logic 1 level. This means that the output will always be at level 1. This identity can be written as $A + \overline{A} = 1$, as shown in Figure 8.4.

**Figure 8.4**
The identity $A + \overline{A} = 1$

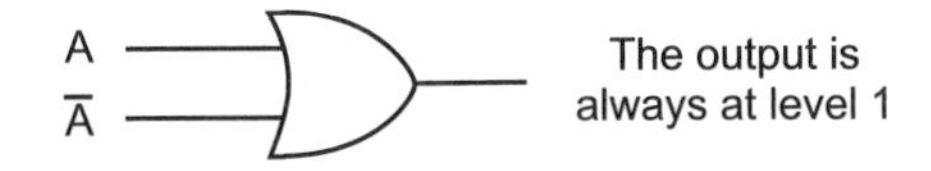


**Unnecessary AND gates**

These follow much the same pattern as we saw with the OR gates – just keep alert to the truth tables.

If we take a 2-input AND gate with one input called A and the other input held at logic 0, the output will always be at logic 0 so the identity could be written as A.0 = 0. This is shown in Figure 8.5.

**Figure 8.5**
The identity A.0 = 0

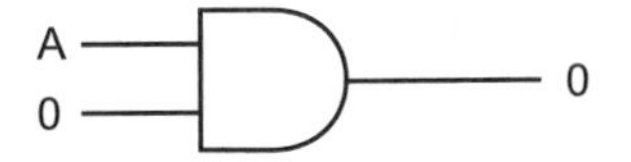


If we build the same circuit but have inputs of A and 1 then the output will always be the same as the A input. We now have the simplification A.1 = A, as in Figure 8.6.

**Figure 8.6**
The identity A.1 = A

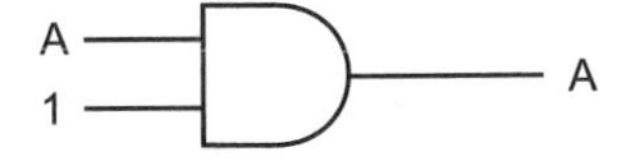


Just like the OR gate, two 0 inputs will result in an 0 at the output, and two 1s will result in a 1 at the output (see Figure 8.7).

**Figure 8.7**
The identity A.A = A

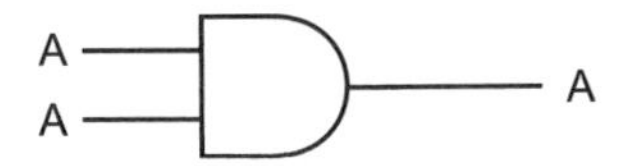


If one input is at level A and the other is at the opposite value, $\overline{A}$, then one of them will always be at a logic 0 level so the AND gate will always produce a 0 output, as in Figure 8.8.

**Figure 8.8**
The identity $A.\overline{A} = 0$

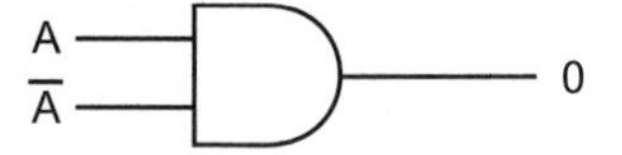


The output is alway 0

**Just as a reminder**

If A is inverted by a NOT gate and then inverted again, we get back to A.

## Summary of identities

1 $A + 0 = A$
2 $A + 1 = 1$
3 $A + A = A$
4 $A + \overline{A} = 1$
5 $A.0 = 0$
6 $A.1 = A$
7 $A.A = A$
8 $A.\overline{A} = 0$
9 $\overline{\overline{A}} = A$

## Using identities in real circuits

If we look at the first identity again, we are saying that the first term ORed with 0 gives an output equal to the first term. This is true whatever the input signal is, whether it is a simple A like we used in our examples or a more complex term.

In Figure 8.9 we used the term $A B \overline{C}$ as the input to the OR gate and, since we know that the identity (some signal) + 0 = (the same signal), we can say

$$A B \overline{C} + 0 = A B \overline{C}$$

and therefore the OR gate can be removed.

This flexibility in what we accept in place of 'A' is an important step to take.

**Figure 8.9**
Using an identity in a real circuit

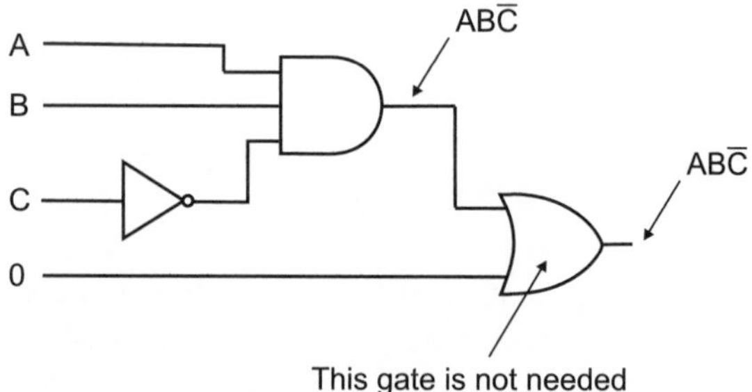

**Example**

Simplify the Boolean expression $A\overline{B}\,\overline{C} + \overline{A\overline{B}\,\overline{C}}$.

This looks complicated until we recognize that the two terms are the same except that the second term has a line across the whole expression showing that it has been inverted. We really have (something) + (same thing inverted). This is just our identity $A + \overline{A}$ in disguise. Since $A + \overline{A} = 1$, it follows that

$$A\overline{B}\,\overline{C} + \overline{A\overline{B}\,\overline{C}} = 1$$

**Example**

Simplify $CK.\overline{\overline{CK}}$

The second term has been inverted twice and we know that inverting something and then inverting it again will get back to the starting point. If we spot a double inversion, always start by cancelling the inversions before trying any other simplifications. Our expression is now CK.CK, which looks like our identity A.A = A. This means that our original expression can be reduced to CK, thus saving two NOT gates and one AND gate.

**Note**: To cancel two inversions, the lines must be the same length. The expression $\overline{A.\overline{B}}$ cannot be simplified. The order in which the bars are added does not matter, so $\overline{AB\overline{\overline{C}}}$ can be simplified to $\overline{ABC}$.

## Commutative law

Do you agree that the two logic circuits in Figure 8.10 are the same?

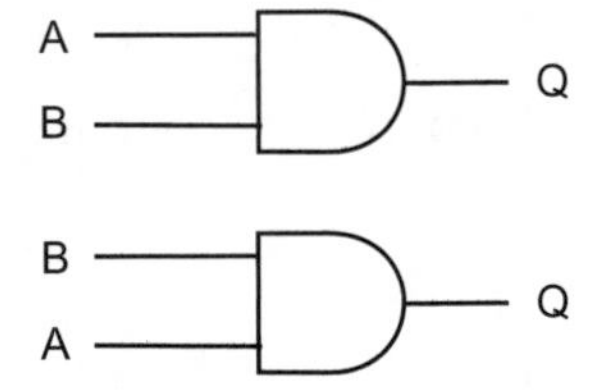


**Figure 8.10**
Are these circuits the same?

Yes, the inputs are just written in a different order. So AB = BA and, of course, in a 3-input gate ABC = ACB = BAC = BCA = CAB = CBA.

We have always known that this applies to addition so 3 + 4 = 4 + 3.

The 'any order' feature of both Boolean and addition is the commutative law. Try some simple sums to see if subtraction, multiplication and division are commutative.

Applying the commutative law to the identities, we can see, for example, 1 + A is the same as A + 1.

In more complex terms, there is much to be said for using the commutative law to re-arrange the terms in alphabetical order. This makes recognition of terms much easier. At first glance, it is easy to miss that ABC, BAC and CBA are the same function.

## Associative law

If we are using 2-input AND gates and we want to AND three terms like A.B.C, then we have to do two of them and then AND the last one to the result.

The circuits in Figure 8.11 show the associative expressions (A.B).C = (A.C).B; we can see that in the first case the A and the B are grouped or associated, and in the second expression it is the A and C that are associated.

**Figure 8.11**

Are these circuits the same?

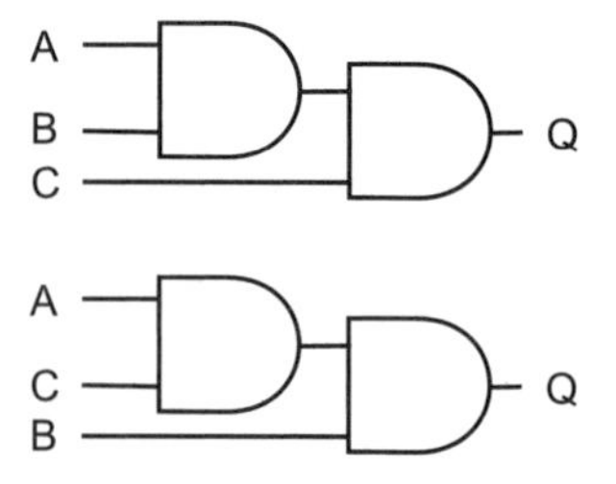


Likewise, OR gates can be used in a similar circuit to give

$$(A + B) + C = (A + C) + B$$

Here are a couple of random examples of the associative law in action.

$$(ABC).D = (DA).(BC)$$

$$(\overline{A} + B + C) + \overline{D} = (\overline{A} + B) + (C + \overline{D})$$

Be careful here. The order does not matter providing the circuit only contains the same type of gate.

However, the order does matter when an expression includes both AND and OR functions as in the expression $\overline{A} + \overline{C}.D$; this is always taken to mean the same as $\overline{A} + (\overline{C}.D)$. If no brackets are used, the AND function is always done first.

**Example**

Simplify $\overline{ABC} + \overline{AB(C)} + ABC = Q$.

First, we can remove the brackets around the letter C to give $\overline{ABC} + \overline{ABC} + ABC = Q$ by associative law.

Now we can combine the first two terms to give $\overline{ABC} + ABC = Q$ by identity number 3.

Now we can finish the simplification: $1 = Q$ by using identity number 4. So none of these gates are needed and the output at Q is held permanently at logic 1. Quite a saving!

**Example**

Simplify $\overline{ACB}.BA(C.1).D = Q$

Simplify to $\overline{ABC}.AB(C.1).D = Q$ (alphabetical order by commutative law – this is optional but helpful).

Then to $\overline{ABC}.AB(C).D = Q$ (identity 6 to change C.1 to C).

To $\overline{ABC}.ABC.D = Q$ (brackets removed by associative law).

$0.D = Q$ (the terms $\overline{ABC}.ABC = 0$ by identity 8).

$0 = Q$ (D removed by identity 5).

This is another worthwhile simplification.

We have seen from these examples that the associative and commutative laws help in the clear layout and understanding of the logic expressions, but so far it is only the identities that actually simplify the circuits.

**Reminders**

1 To be slick at using Boolean algebra for simplification it is very important that we know the identities and can spot them when they arise.
2 Identities always involve the following:
   (a) Two identical terms like $A\overline{B}C$ and $A\overline{B}C$.
   (b) An inverted but otherwise identical terms like $\overline{AB\overline{C}}$ and $AB\overline{C}$.
   (c) A logic level 0 or 1.

## Distributive law

This is the last Boolean law and, like the identities, it is able to reduce the number of gates in a circuit. The logic diagram and, of course, the Boolean expression always contains a mixture of both AND gates and OR gates.

Have a look at the logic diagram in Figure 8.12. We can see that it performs the function $Q = (A.B) + (A.C)$.

How can we say this in words? We can say that in this circuit (A is ANDed with B) OR (A is ANDed with C). With a slight change in emphasis, we could also say that A is ANDed with B OR C.

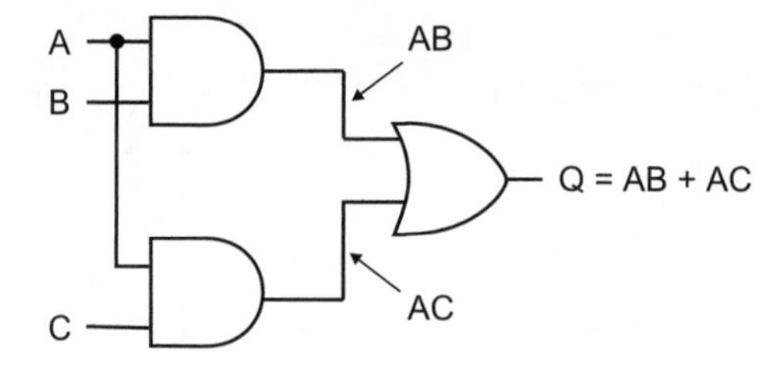


**Figure 8.12**
Before applying the distributive law

Now, if we write this in Boolean, we say A is ANDed with (B + C) or more simply that Q = A.(B + C).

We have now described the function as Q = (A + B).(A + C) and also as Q = A.(B + C). We can see in Figure 8.13 that we have now used only two gates instead of three and, if we draw up the truth tables, they would be identical.

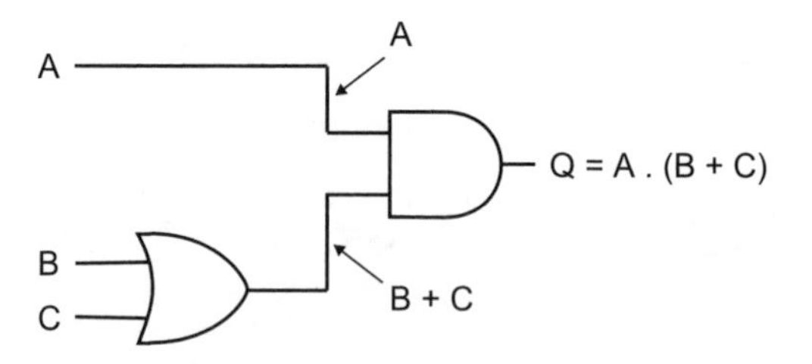


**Figure 8.13**
Another gate saved

If we take the second expression, Q = A.(B + C) we say that the A. (A AND) has been applied to both of the other terms. Technically we say that the 'A.' has been 'distributed over' the other two terms – hence the name 'distributive law'.

If we have a different expression like Q = A + (B.C), we can distribute the 'A OR' over the other two terms, giving the logic diagrams in Figure 8.14.

Q = A + ( B . C )

Q = (A + B) . (A + C)

There are three things to note about the distributive law:

1 It does not follow the rules of 'normal' algebra.
2 It is only used when both AND and OR gates are used.
3 It can be used to expand or reduce Boolean expressions.

**Figure 8.14**
Another distributive example

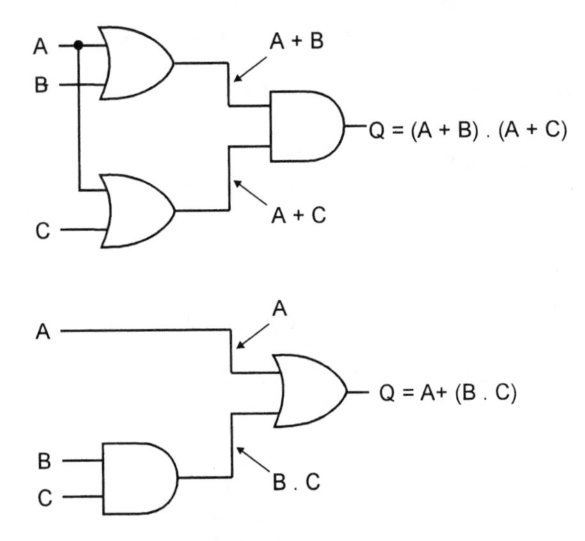


**Example**

Simplify the expression $\overline{A}BC\overline{D} + \overline{A}BCD = Q$

Always start by checking for identities. In this case there are none since there are only two terms, and they are neither the same nor the inverse of each other. This means that identities are non-starters, so we must move on by trying the distributive rule.

Look for the common parts of the two expressions. The $\overline{A}$ is common, and so is the B and C, so the common part ABC can be filtered out by the distributive law to give $\overline{A}BC(\overline{D} + D) = Q$.

Now we instantly spot an identity $\overline{D} + D = 1$, so it becomes $\overline{A}BC.1 = Q$, and by another identity we have ABC = Q.

This is as far as we can go, but this is useful in that it has shown that the input D is not used. Sometimes there are surprising practical consequences such that whole sections of a design are proved redundant.

**Example**

Simplify $\overline{AB}.(AB + C\overline{E}) = Q$.

Check for identities. There are none, so apply the distributive law to give

$$(\overline{AB}.AB) + (\overline{AB}\ \overline{CE}) = Q$$

Notice that we cannot use a single bar over all the terms.

Now what? Let's sort out the first bracket with an identity.

$$(0) + \overline{AB}\,\overline{CE} = Q \text{ (using identity 8.)}$$

Use another identity to remove the 0 term:

$$\overline{AB}\,\overline{CE} = Q$$

## Absorption laws

These are not really new laws at all, but are merely a summary of results that we have already seen by using the distributive laws and the identities.

They are used to remove or 'absorb' a term. There are four of them:

1 $A + AB = A$
2 $A + \overline{A}B = A + B$
3 $A(A + B) = A$
4 $A(\overline{A} + B) = AB$

If we can spot the absorption patterns we can save several steps in a simplification but if not, it doesn't matter; all of these simplifications can always be done by using the distributive law and identities. They are helpful but not essential.

As with identities, letters like A, B etc. can be used to represent groups of letters providing they fit into the same general pattern.

For example, taking the absorption law $A + AB = A$ and comparing it with the expression $FAC\overline{B} + A\overline{B}CFE$. We start by putting it in alphabetical order by using the commutative law:

$$A\overline{B}CF + A\overline{B}CEF$$

This is an absorption law in disguise. The first term is repeated with only an additional E so, using the commutative again, we can make it look like the standard absorption law:

$$A\overline{B}CF + A\overline{B}CFE$$

The relationship with the standard form of the absorption law is shown in Fig. 8.15.

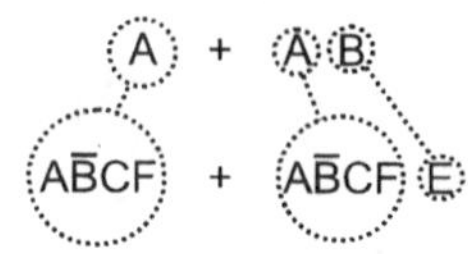


**Figure 8.15**
Relationship with the standard form of the absorption law

## Real life simplification

Having looked at all the laws of Boolean, we have the tools necessary to simplify any expression that can be simplified.

The normal method is to use the identities and the distributive law over and over again. The whole process is very much like a time-absorbing game, where the rules are simple but the degree of ingenuity and skill pays off in money savings and an inner glow of satisfaction.

There is no definite procedure that will 'solve' all simplifications. It is a matter of following the rules and trying a method of attack. If it goes wrong, don't be downhearted – everyone takes the wrong path from time to time.

## General method

1 Look for identities and simplify using commutative and associative laws as we go.
2 Apply the distributive rule.
3 Look for an identity and simplify.
4 Apply the distributive rule.
5 Carry on like this until something tells us to stop – we may have completed the simplification, we may be back where we started or perhaps nothing useful appears to be happening.

If we get back to where we started, take comfort in the knowledge that at least we must have applied all the rules correctly. So getting back to the start is really an achievement in itself. We just start again by a different route.

How do we know if our simplification is correct? We can always check a result by drawing up the truth table of the original expression and the final result.

### Example

This is a real example, which shows all my working (including the mistakes) so we can see the sort of thing that really happens.

Simplify A + AB = Q.

Check for identities. There aren't any. Note that we cannot use the A + A as an identity since the second A is only part of the expression AB. We must use either the whole of an expression or none of it. As there are no identities, we either abandon the task as hopeless or we apply the distributive law:

$(A + A)(A + B) = Q$ (by distributive law)

Check for identities. We see A + A = A, so we have:

$(A)(A + B)$

There are no more identities, so we use the distributive law to expand the expression by applying A to each term in the other bracket:

$$(A.A) + (A.B) = Q$$

Look for identities. The first bracket contains A.A, which can be reduced to A.

$$A + (A.B) = Q$$

which is where we came in . . .

We have gone in a complete circle. Never mind, let's try something else.

There is another identity A.1 = A, so we could apply this to the first A in the expression. We now have:

$$(A.1) + (A.B) = Q$$

We can see that both terms include the term A so the common factor, A, can be separated out by using the distributive law. This results in

$$A.(1 + B) = Q$$

Inside the bracket we have the identity 1 + B, which can be replaced by 1 to give:

$$A.1 = Q$$

This is another identity which reduces to A = Q, so we have finally reduced A + AB to just A so the input B played no part in the outcome.

## What we have learned

With experience it is possible to 'look ahead' a little to avoid taking the wrong route too often, but even so we must expect a false start from time to time.

The general rule is that if you are on the right track, things will be looking distinctly better after about five lines of work. Remember, as illustrated in the last example, the distributive law can be used to expand an expression as well as to reduce it.

Make notes as you go along. This allows easy checking if an error is suspected but when doing it for yourself, the notes can be abbreviated.

### Example

Show that $\overline{A}B\overline{C} + A\overline{B}C + AB\overline{C} + ABC = B\overline{C} + AC$.

In the answer, we have the term $B\overline{C}$. Now this is part of the first and third terms. The second term in the answer is AC, and this occurs in the second and fourth terms. We may find it easier if we rewrite the question in a different order to bring these terms together.

$\overline{A}B\overline{C} + AB\overline{C} + A\overline{B}C + ABC = B\overline{C} + AC$ commutative law.

Now to attack the first two to extract the $B\overline{C}$ term:

$B\overline{C}(\overline{A} + A) + A\overline{B}C + ABC = B\overline{C} + AC$ distributive law.

$B\overline{C}(1) + A\overline{B}C + ABC = B\overline{C} + AC$ identity.

$B\overline{C} + A\overline{B}C + ABC = B\overline{C} + AC$ identity.

We have finished the first term of the answer – now for the second one. The same process will be followed.

$B\overline{C} + AC(\overline{B} + B) = B\overline{C} + AC$ distributive law.

$B\overline{C} + AC(1) = B\overline{C} + AC$ identity.

$B\overline{C} + AC = B\overline{C} + AC$ identity.

Finished!

**Example**

Show that $PQS + RP + P\overline{Q}\overline{R}S = P(S + R)$.

$PS(Q + \overline{Q}\overline{R}) + RP = P(S + R)$ distributive law applied to first and third terms.

$PS(Q + \overline{R} + RP = P(S + R)$ absorption law.

$PSQ + PS\overline{R} + RP = P(S + R)$ distributive law.

$PSQ + P(S\overline{R} + R) = P(S + R)$ distributive law.

$PSQ + P(S + R) = P(S + R)$ absorption law.

$PSQ + PS + PR = P(S + R)$ distributive law.

$PSQ + (PS.1) + PR = P(S + R)$ identity.

$PS(Q + 1) + PR = P(S + R)$ distributive law.

$PS.1 + PR = P(S + R)$ identity.

$PS + PR = P(S + R)$ identity.

$P(S + R) = P(S + R)$ distributive.

## Summary of Boolean laws and identities

### Identities

1 $A + 0 = A$
2 $A + 1 = 1$
3 $A + A = A$
4 $A + \overline{A} = 1$
5 $A.0 = 0$
6 $A.1 = A$
7 $A.A = A$
8 $A.\overline{A} = 0$
9 $\overline{\overline{A}} = A$

### Commutative law

$A.B = B.A$
$A + B = B + A$

### Associative law

$A + (B + C) = (A + B) + C$
$A.(B.C) = (A.B).C$

### Distributive law

$A + (B.C) = (A + B)(A + C)$
$A.(B + C) = (A.B) + (A.C)$

### Absorption law

$A + AB = A$
$A + \overline{A}B = A + B$
$A(A + B) = A$
$A(\overline{A} + B) = AB$

## Why does Boolean algebra use + for OR and . for AND?

Boolean algebra originated from a study of probability, which is the method used for predicting the likelihood of an event occurring in the future. A probability of 0 means the event is not possible or has a zero probability, whereas a probability of 1 is a certainty.

Let's assume that we have four cards marked with the letters A, B, C, and D all face down on a table.

Selecting a card at random, the probability of choosing the 'A' card is 1 in 4, or $\frac{1}{4}$. This is written as probability of A = $\frac{1}{4}$ or P(A) = $\frac{1}{4}$. If we replace the card, shuffle them and again choose a card at random, then the probability of the chosen card being B will also be $\frac{1}{4}$. What is the probability of the two cards selected being A AND B? To find this probability we have to multiply the two individual probabilities, so the answer is $\frac{1}{4} \times \frac{1}{4}$ or $\frac{1}{4}.\frac{1}{4}$ = 1/16.

This situation would be written as shown in Figure 8.16a.

**Figure 8.16(a)**
Probabilities. See how the 'AND' function appears in the formula as a dot (.)

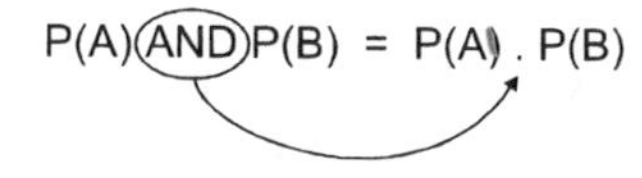


If we had asked a different question and said 'what are the chances of getting an A OR a B?' the probability would be much greater. In fact it would be the result of adding the two individual probabilities (Figure 8.16b).

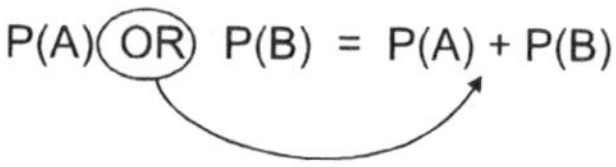

**Figure 8.16(b)**
Probabilities. Now see how the 'OR' function appears in the formula as a + symbol

P(A) OR P(B) = P(A) + P(B)

## Quiz time 8

In each case, choose the best option.

**1 In arithmetic, the commutative law applies to:**

(a) addition and multiplication.
(b) addition and subtraction,
(c) multiplication and division.
(d) subtraction and division.

**2 Which of these is NOT an identity?**

(a) H + 0 = H
(b) R.R = 1
(c) W + 1 = 1
(d) (G)(0) = 0

**3 A reduction in the number of gates can be achieved by using the:**

(a) identities or the commutative law.
(b) the commutative and the associative laws.
(c) identities or the distributive law.
(d) identities but none of the other laws.

**4 The Boolean expression $A + \overline{A}B$ is equivalent to:**

(a) AB
(b) A + B
(c) 1 + AB
(d) $\overline{AB}$

**5 The Boolean expression $ABC + B\overline{C} + ABC$ can be simplified to:**

(a) AB
(b) 1
(c) C
(d) $\overline{C}$

# 9

# Simplifying – by Karnaugh maps

There have been many attempts to find an easy way of simplifying logic circuits. We have already looked at Boolean algebra, and now we will consider the contribution offered by Maurice Karnaugh (pronounced KAR-NO), an American physicist.

Mr Karnaugh has given us the Karnaugh map. This is a graphical means of simplifying a Boolean for people who don't like or don't know the Boolean laws.

## Advantages of Karnaugh mapping

1. Boolean laws are not needed.
2. The best simplification is always obtained, whereas with Boolean we often obtain several different solutions and are still not sure if there is a better one.
3. A Karnaugh map can be used to decide between alternative Boolean answers.
4. We can work directly from a truth table.

## Disadvantages of Karnaugh mapping

1. De Morgan's laws may (sometimes) still be needed unless we are working directly from a truth table.
2. People who are really slick with Boolean simplifications complain about this method being slower.

3 In reality, Karnaugh mapping is only used for terms containing three or four different letters. For cases involving less than three, Boolean or the identities are so much quicker. Although Karnaugh mapping can be used for up to eight letters it becomes rapidly more difficult over four letters, so we normally go back to Boolean. Nowadays, larger complicated expressions enlist the help of computers.

## Why are De Morgan's laws used?

To use a Karnaugh map, we should first ensure that the Boolean expression is written in this form: F = ABC + AB + ABC . . ., that is, ANDed terms which are then ORed together. The technical name for this format is the 'minterm' or the 'sum-of-products', and is the normal form of the Boolean expression when we have started with a truth table or a practical design.

**Note**: It is possible to use Karnaugh maps with 'maxterms' like F = (A + B + C).(A + B).(A + B + C) in which ORed terms are ANDed together, but we will find it a lot easier to use De Morgan's laws and change it to a minterm first.

## Drawing a Karnaugh map

The Karnaugh map is always drawn as a rectangular grid. The number of squares in the grid is equal to the number of lines in the truth table.

### Example

How many squares would there be on a Karnaugh map if the Boolean expression contained three terms?

Answer:

You may recall that the number of lines in a truth table is $2^n$ where n is the number of letters used in the Boolean expression, so in this case we would have $2^3$ lines in the truth table and therefore also $2^3 = 8$ squares on the Karnaugh map.

## The shape of the map

The shape is always rectangular, so with 8 square we have a choice of $2 \times 4$ or $4 \times 2$ as in Figure 9.1. It really doesn't matter which you use, your final answer will be exactly the same. The squares are numbered using binary numbers in a special order called the 'Gray' code.

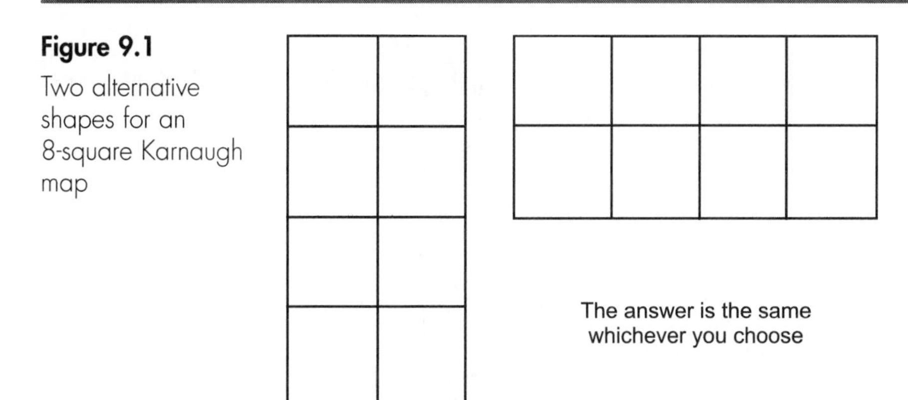


**Figure 9.1**
Two alternative shapes for an 8-square Karnaugh map

## The Gray code

This is often preferred to straight binary when counting is carried out by mechanical or electronic means. By counting in a slightly different order we do not run the risk of an accidental error caused by the switching process. To see how this may occur, let's start counting up in binary:

000
001
010

Yes, 0, 1, 2, no problem.

When the count changed from 001 to 010 the last two bits had to change in value. The right-hand bit changed from 1 to 0, and at the same time the next bit changed from 0 to 1.

Consider what would happen if the second bit changed slightly faster than the last bit. Just for a moment we would have the situation where the count would actually be 011. This might only occur for a very short time, probably too fast for us to see, but still slow enough to be spotted by a digital circuit.

We would expect a count of 000, 001, 010, or 0, 1, 2, but we could actually have 000, 001, 011, 010, or 0, 1, 3, 2. Going from a count such as 0111 to 1000 could throw up all sorts of possibilities.

This problem may occur each time there is a change by more than a single bit from one count to the next. This gives us the clue to the design of the Gray code. This code never allows more than a single bit to change in a single count. The count may then be late or early, but it can never throw up a false count.

The Gray code rule is to start at zero using however many bits we require, then, on the first count, change the right-hand bit. At each succeeding count, change the bit furthest to the right unless it would create a number that has already been used, in which case simply move one bit further to the left and try changing that one.

### An example using four bits

Start with 0000.

On the first count change the right-hand bit to give 0001.

On the next count we cannot change the right-hand bit because it would change the number back to 0000, which has already been used, so we move one bit to the left and change that one. This gives a value of 0011.

On the next count we can change the right-hand bit because it would result in an unused number of 0010.

On the next count we cannot change the right-hand bit because it would change the number to 0011, which has been used, so instead we move one bit to the left and change that one. This again gives a number – 0000 – that has been used. So we have to move one more bit towards the left and change the next bit. This would give 0110, which has not been used so is acceptable.

If we keep counting by the Gray code it will eventually run out of possible changes and will have to restart from 0000. Any code that naturally recycles like this is called a cyclic code. Notice how the 0s and 1s make a regular pattern as we read down the columns – compare the binary pattern to the Gray pattern.

| Decimal | Binary | Gray code |
|---|---|---|
| 0 | 0000 | 0000 |
| 1 | 0001 | 0001 |
| 2 | 0010 | 0011 |
| 3 | 0011 | 0010 |
| 4 | 0100 | 0110 |
| 5 | 0101 | 0111 |
| 6 | 0110 | 0101 |
| 7 | 0111 | 0100 |
| 8 | 1000 | 1100 |
| 9 | 1001 | 1101 |
| 10 | 1010 | 1111 |
| 11 | 1011 | 1110 |
| 12 | 1100 | 1010 |
| 13 | 1101 | 1011 |
| 14 | 1110 | 1001 |
| 15 | 1111 | 1000 |

## Numbering the Karnaugh map

Having drawn out the Karnaugh map, the next step is to number each square across the top and also down the left-hand side starting from the top left-hand corner. Using our Karnaugh map with eight squares, we should number it as shown in Figure 9.2.

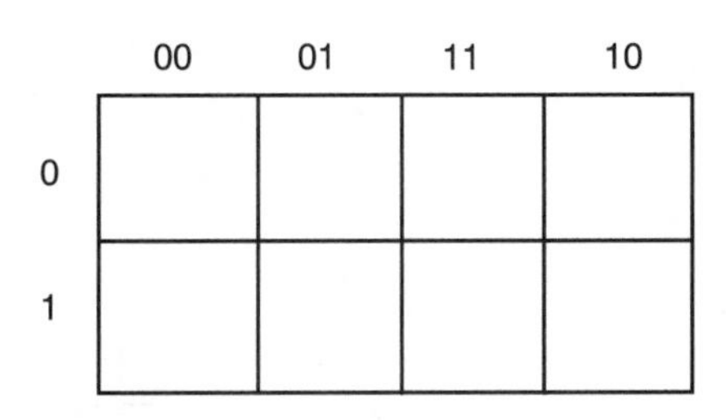


**Figure 9.2**
Numbering the squares

Boolean expressions use letters, of course. If we allocate the first two letters to the sequence across the top and the third letter to the sequence down the side we could use the letters AB and C (Figure 9.3).

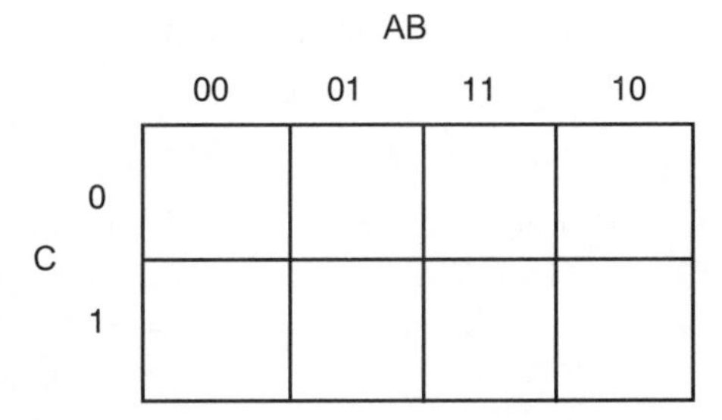


**Figure 9.3**
Adding the letters

The top left-hand square has the co-ordinates AB = 00 and C = 0. We always take A as equal to 1, so this makes $\overline{A}$ equal to zero and we could describe this square as A = 0, B = 0, C = 0 or $\overline{A}\,\overline{B}\,\overline{C}$.

The square immediately under this one has the co-ordinates A = 0, B = 0 and C = 1, so this would become $\overline{A}\,\overline{B}\,C$ in Boolean (Figure 9.4).

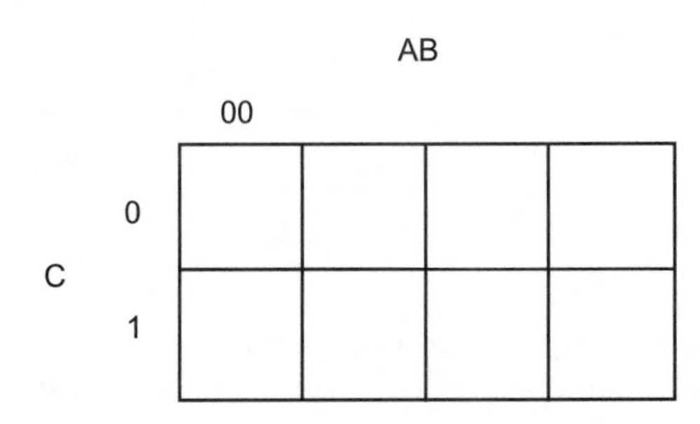


**Figure 9.4**
Describing the squares

**Example**

In Figure 9.4, complete the square numbering and write the Boolean equivalent of each square.

Answer:

We don't normally write in these Boolean descriptions on the map, but in this case it is worthwhile to make quite sure that we are happy with the Karnaugh map. The results are shown in Figure 9.5.

**Figure 9.5**
The full description

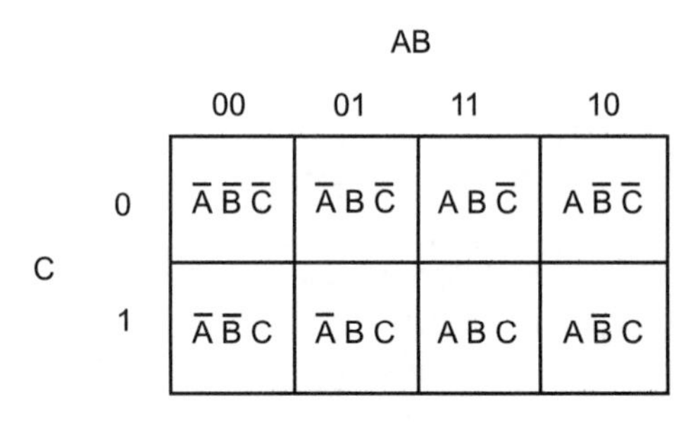


## Simplification by Karnaugh map

The easiest way to see the Karnaugh map in operation is to use it to solve something where we already know the answer.

Here's one we did earlier. Simplify the Boolean expression $Q = AB\overline{C} + ABC$. From our work in Chapter 8 we may be able to see that this can be simplified to $Q = AB$ by using the absorption law. If you've forgotten, it doesn't matter – we can do it without Boolean.

Step 1: Draw the basic map with the numbers and letters on the axes as in Figure 9.6.

**Figure 9.6**
The starting point

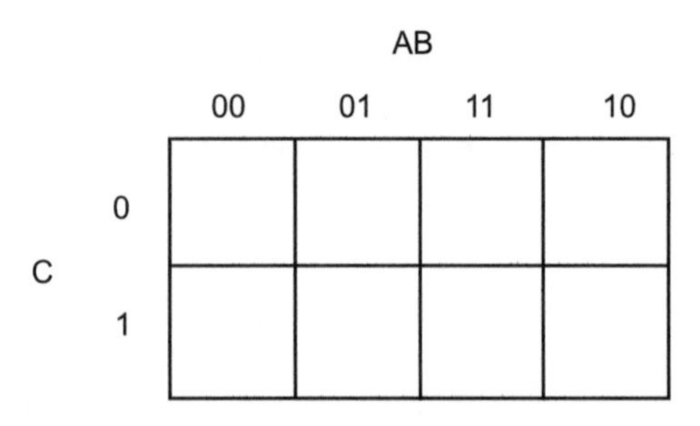


Step 2: Each Boolean term refers to a square on the map. In each square referred to, insert a number '1' in the square. See Figure 9.7.

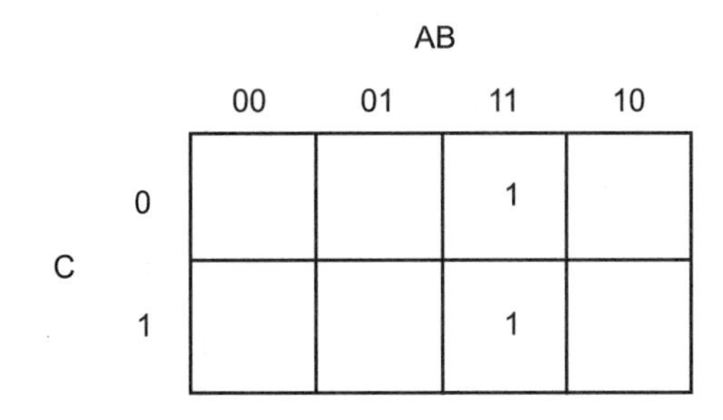


**Figure 9.7**
Add a '1' for each Boolean term

Step 3: Group together the largest number of '1' cells that form a rectangle or a square on the map in which the number of squares is a power of 2, i.e. 1, 2, 4, 8, 16, 32 etc. In this case there is only one possibility. This grouping is seen in Figure 9.8.

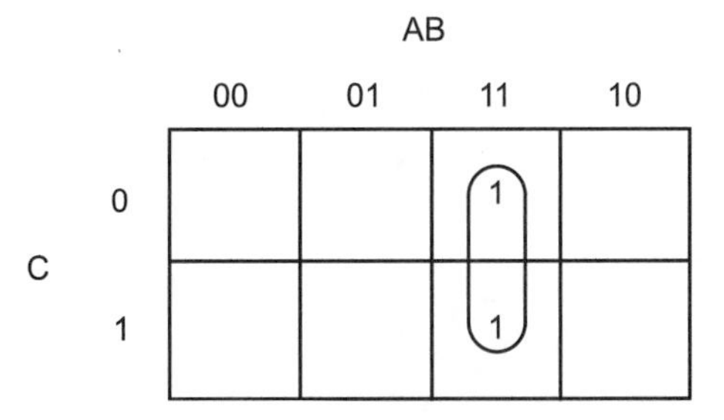


**Figure 9.8**
Group the boxes together

Step 4: This is where we produce our answer by writing the Boolean expression for the group of squares.

To do this, we consider the group as a whole and see what letters are needed to describe it. In our example both squares have A = 1 and B = 1 since both squares are in the same column, so if we were asked to shade in the whole area in which A = 1 and B = 1 we would include both of the encircled squares. The C information does not matter since part of the circled pair has C = 1 and part has C = 0. We just ignore any letter that changes its mind.

The group only needs the A and B terms to describe the whole group so the simplification is Q = AB, just as we found with Boolean.

**Example**

Use a Karnaugh map to simplify the Boolean expression $F = \overline{A}\,\overline{B} + \overline{A}\,\overline{B}C + \overline{A}BC + \overline{A}B\overline{C}$. We start in the same way as before.

Step 1: Draw the map adding letters and numbers as in Figure 9.9.

**Figure 9.9**
The first step

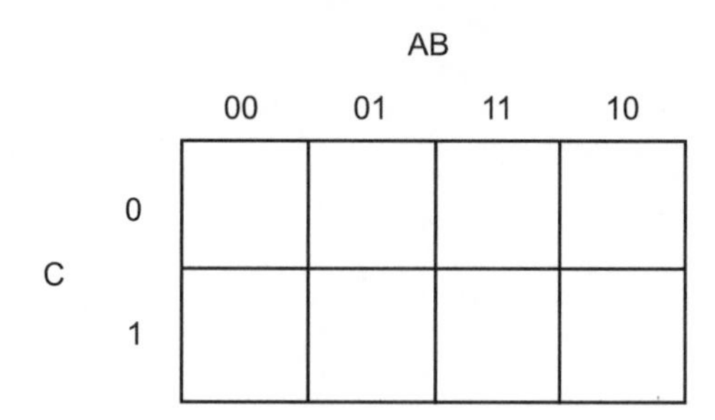


Step 2: Enter each term by putting a '1' in the relevant squares. There are two new points here. First, the term $\overline{A}\,\overline{B}$ describes both squares in the left-hand column in much the same way as we saw in the previous example. The second term $\overline{A}\,\overline{B}\,C$ falls in the bottom left-hand square, which has already been marked. This doesn't matter – just mark any square once. In fact this shows that the term $\overline{A}\,\overline{B}\,C$ was not needed, so we have achieved some simplification already.

The third term $\overline{A}\,B\,C$ will account for the bottom square in the second column, and $\overline{A}\,B\,\overline{C}$ will account for the other square in this same column. Have a look at Figure 9.10.

**Figure 9.10**
Adding the four terms

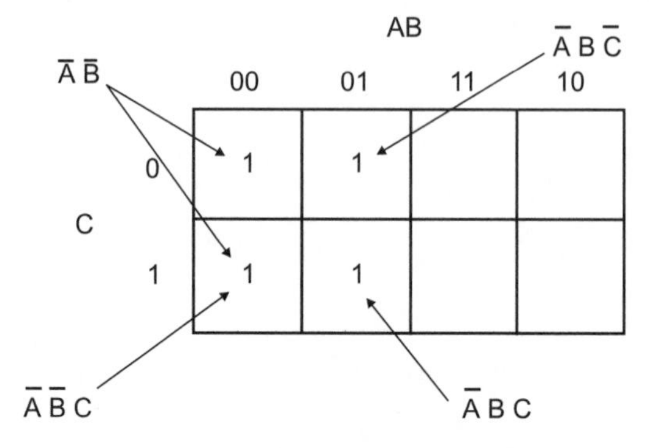


Step 3: Group the cells using the largest possible rectangle or square of sizes 1, 2, 4, 8,16, 32. In this case these four terms conveniently form a 2 × 2 square. Why not two pairs like in the last example? This would be 'a' simplification, but not the best one. The larger the group, the simpler the result.

Step 4: Writing the Boolean. Looking at the square in Figure 9.11 and, taking each letter in turn, we see that both columns include $\overline{A}$. Both B and C change their mind and can be ignored. This means that the whole square can be described as $\overline{A}$.

**Figure 9.11**
The best simplification

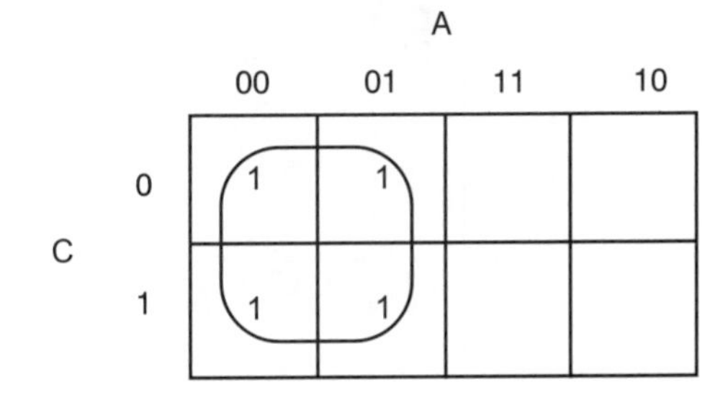


This is a really worthwhile simplification; we have $F = \overline{A}\,\overline{B} + \overline{A}B\overline{C} + \overline{A}BC + \overline{A}\,\overline{B}C = \overline{A}$, and the whole digital circuit can be thrown away and replaced by a single inverter on the A input. If you feel doubtful of this you can spend a few minutes writing out the truth table or playing with the Boolean algebra laws.

## Grouping

The Karnaugh map is continuous. If we move across the map from left to right, we leave the right-hand edge and reappear on the left-hand edge just like the map in Figure 9.12. The implication of this wrapping is that, in Figure 9.13, square A and B are adjacent to one another. The same goes for C and D so, curiously enough, the four squares A, B, C and D form a square.

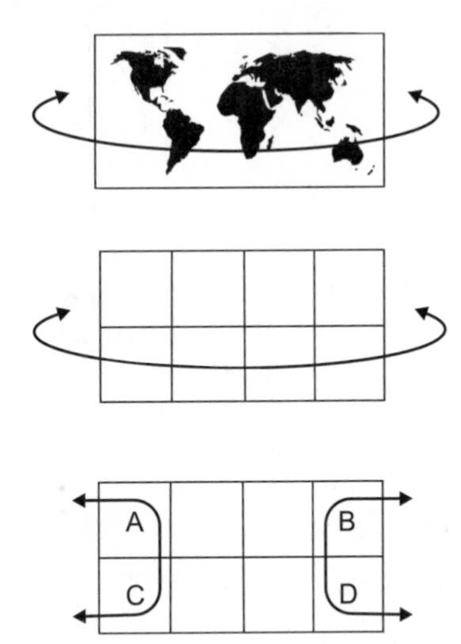


**Figure 9.12**
Wrapping the world and the Karnaugh map

**Figure 9.13**
An unexpected square

**Example**

Write the simplified Boolean expression for each of these Karnaugh maps (Figure 9.14).

**Figure 9.14**

Find the simplest Boolean form of these maps

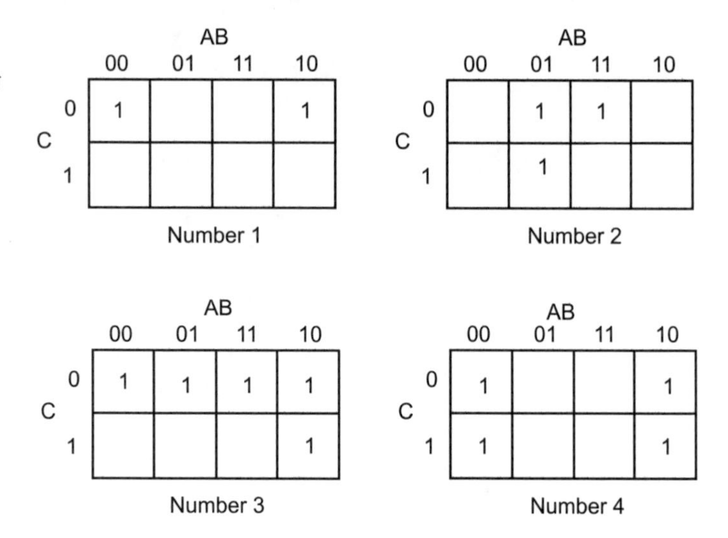


Answers:

**Map number 1**:

Owing to the way the map wraps, the top left-hand corner is adjacent to the top right-hand corner and so we can group them as in Figure 9.15.

**Figure 9.15**

These squares can be grouped together

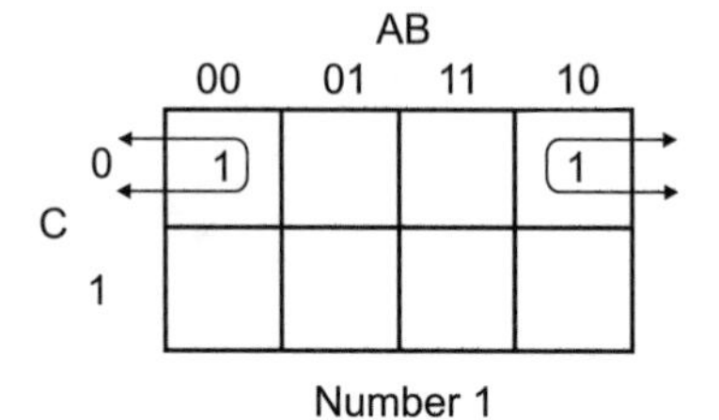


We just have to look at the A, B and C terms to see which are of fixed value and which are changing. First, A: in the top left-hand corner A = 0 and in the other square A = 1. Since A changes its value we can just ignore it so the final simplification will not contain any A term. Now for B. In both squares B = 0, so it will have a value of $\overline{B}$. Finally, both active squares are in the same column in which the C term = 0, so C has the value of C.

The final simplification is $\overline{B}\,\overline{C}$.

**Map number 2**:

In this map we have three ways of grouping the three active squares. We can group any two of them and leave one square all on its own, or we can use one of the squares twice. These options are shown in Figure 9.16.

**Figure 9.16**
We often get more than one option

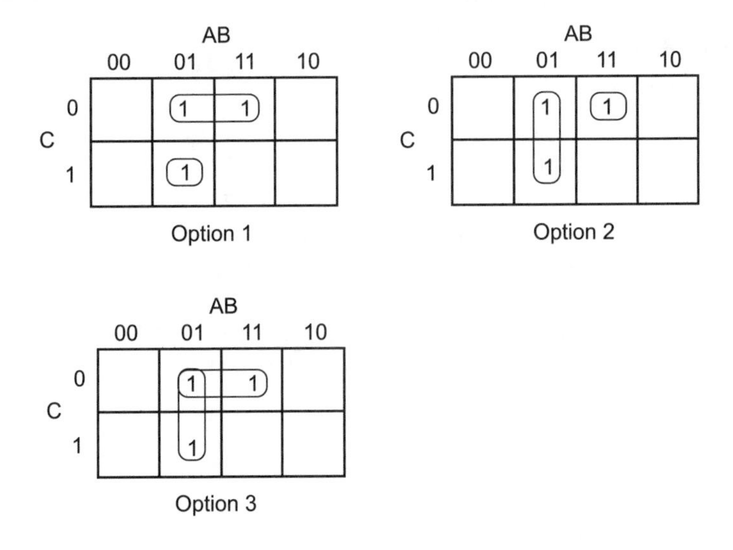


The first option obtained by grouping the two on the top line is $B\overline{C}$ because the A term is switching from 0 to 1 and is therefore ignored. The square on its own is A = 0, B = 1, C = 1, which gives $\overline{A}BC$.

This gives $B\overline{C} + \overline{A}BC$.

The second option is to group the two active squares in the A = 0, B = 1 column to give $\overline{A}B$, then deal with the little lonely square, which will be called $AB\overline{C}$.

The third, and preferred, option is to group it into two groups of two. We have already found the description of the groups in the previous options, so this result will be $\overline{A}B + B\overline{C}$.

Why is this final option preferred? The last one is best because of the gates needed to build it. The two terms are ORed together so we need one 2-input OR gate and the groups only need two 2-input AND gates. This is a single chip, since we get four 2-input AND gates in one integrated circuit. The other options would need an OR gate, a 2-input AND gate and a 3-input AND gate – three chips in all. This would mean a physically larger circuit to build and additional power supplies.

When grouping, remember always to group the largest number of squares (2, 4, 8, 16) together, even if this means using several squares more than once.

**Map number 3**:

Looking at this map, we can see immediately that the biggest group is the four terms on the top line and there is a group of two in the right-hand column. Have a look at Figure 9.17.

**Figure 9.17**
Always use big groups if possible

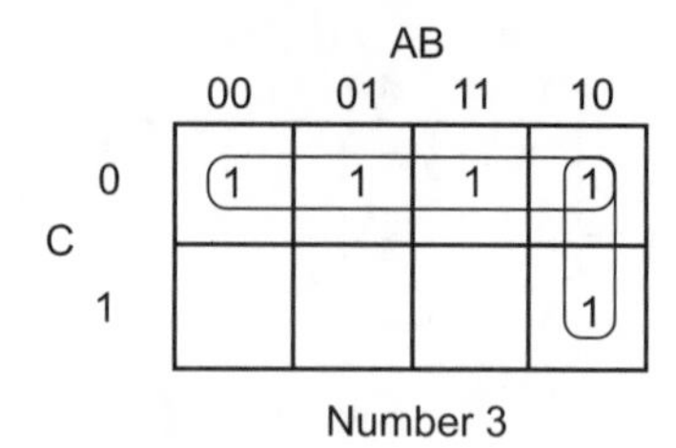


In the four squares across the top, A changes from 0 to 1, B can't make up its mind either, but C is definitely equal to 0. These four squares are therefore described simply as $\overline{C}$.

The two in the last column have A = 1, B = 0 and C is not interested, so this pair is written as $A\overline{B}$.

The final simplification is $\overline{C} + A\overline{B}$.

**Map number 4**:

In this last map, the wrapping will mean that these four squares will form a square (Figure 9.18). Looking at the letters, A and C are changing values so they drop out, which only leaves B at a value of 0. The final simplification is therefore just $\overline{B}$.

**Figure 9.18**
These make a square

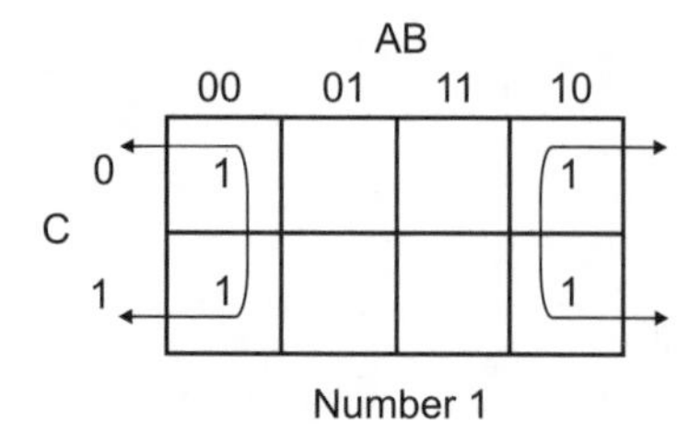


This is worth looking at for another moment. The original unsimplified logic circuit would be $\overline{A}\,\overline{B}\,\overline{C} + A\overline{B}\,\overline{C} + \overline{A}\,\overline{B}C + A\overline{B}C$, which would need a 4-input OR gate and four 3-input AND gates, all of which can be boiled down to one single NOT gate. Very useful.

## Drawing a Karnaugh map for 4-inputs

Using four letter terms, the truth table and the Karnaugh map have 16 entries instead of the previous eight. We now draw the Karnaugh map as a 4 × 4 square, as in Figure 9.19. The letters and the numbering are just a simple extension of the previous map – still Gray code.

Reading the map is much as before. For example, the bottom right-hand corner square would be read as $A\overline{B}C\overline{D}$.

**Figure 9.19**
The four-term Karnaugh map

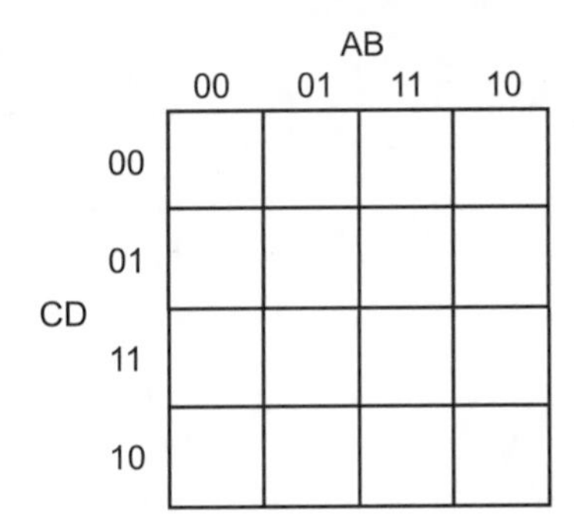


In the 3-input Karnaugh map, we saw that it wrapped round in the horizontal direction. The 4-input map is very similar except that it wraps in the vertical direction as well. In Figure 9.20, the square 'a' is next to square 'b' by wrapping the map horizontally, but it is also next to square 'c' by vertical wrapping. Similarly, 'c' is next to 'd' by horizontal wrapping.

**Figure 9.20**
It now wraps both ways

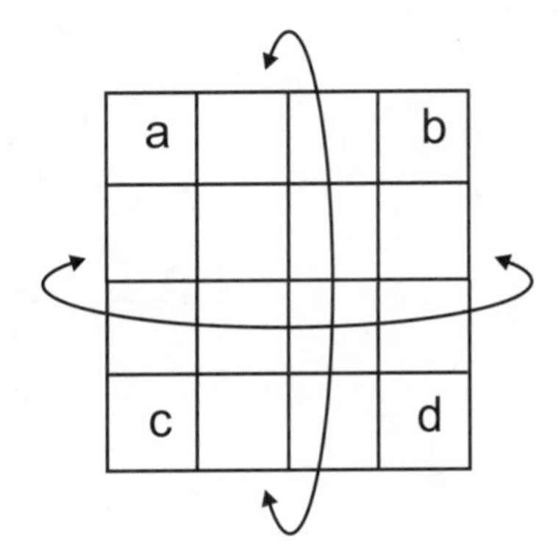


This means that the top row is next to the bottom row as well as the left-hand column being next to the right-hand column. This throws up the curious result that the four squares in the corners are actually adjacent to each other and can form another square.

It probably makes your head hurt when you think about this, but an example may help.

**Example**

Use a Karnaugh map to simplify the following Boolean expression

$$\overline{A}\,\overline{B}\,\overline{C} + \overline{A}\,B\,\overline{C}\,D + A\,\overline{B}\,\overline{C}\,D + \overline{A}\,\overline{B}\,C\,\overline{D} + \overline{B}\,\overline{C}\,D + A\,\overline{B}\,C\,\overline{D}$$

Answer:

Luckily, this is easier than it looks. It just follows the same sequence of steps that we have used in the previous examples.

**Figure 9.21**

Remember to use Gray code

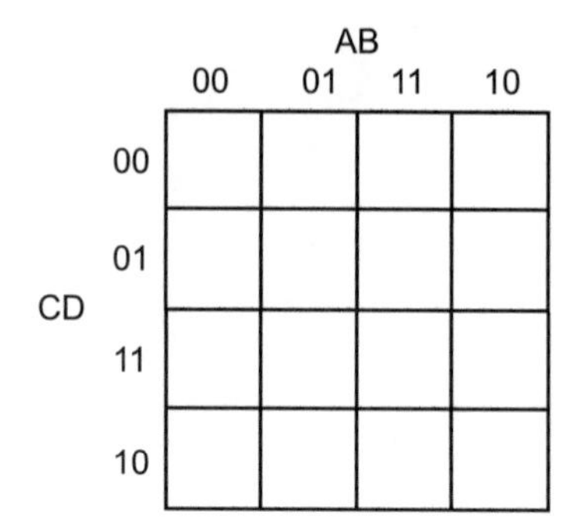


Step 1: Prepare the map as shown in Figure 9.21.

Step 2: Enter the terms. Follow the normal procedure, but remember that some terms can use the same squares. The final result is shown in Figure 9.22.

**Figure 9.22**

Don't worry about duplicates

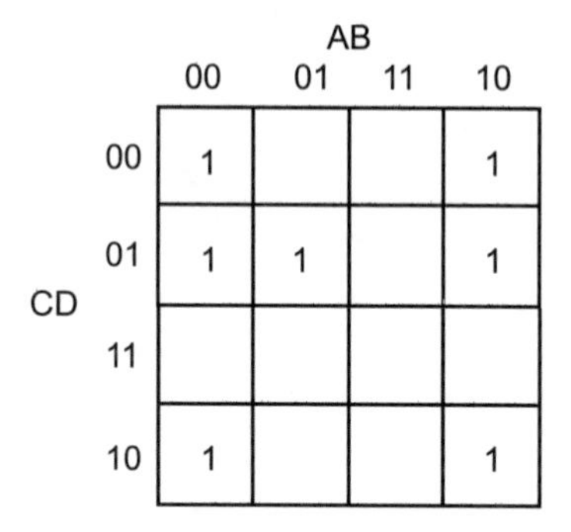


It doesn't matter that some squares are used by more than one term.

Step 3: Group the cells as in Figure 9.23. Notice that the four corner squares are grouped, since the wrapping effect has brought them together.

Step 4: Write the Boolean expression for each of the groups. The four corners are $\overline{B}\,\overline{D}$ because B and D stay at a zero state but A and C change states.

The group of two in the top right-hand corner is $A\overline{B}\,\overline{C}$ and the other pair is $A\overline{C}\,\overline{D}$.

This would give a final simplification of $Q = \overline{B}\,\overline{D} + A\overline{B}\,\overline{C} + \overline{A}\,\overline{C}D$.

**Figure 9.23**
The bigger the better

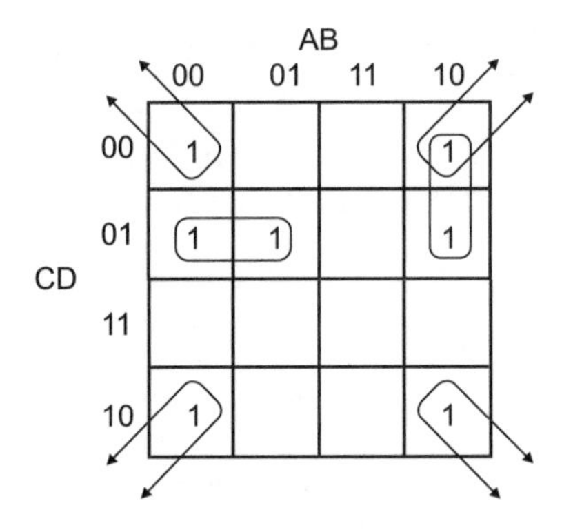


Group them into 16s, 8s, 4s, 2s or 1s.
The larger the group, the better the simplification.

**Notes**:

The top right-hand square has been used in two different groups – it is part of the group of four and is also used as part of the group of two. Once again, this doesn't matter and helps to give the best simplification.

We could have found an alternative simplification by a different choice of grouping as shown in Figure 9.24, which would give the final result as $Q = \overline{B}\,\overline{D} + \overline{B}\,\overline{C}D + \overline{A}\,\overline{C}D$. There is, in this case, more than one possible simplification. This is often the case with digital circuits. We can use whichever we prefer.

**Figure 9.24**
There can be alternatives

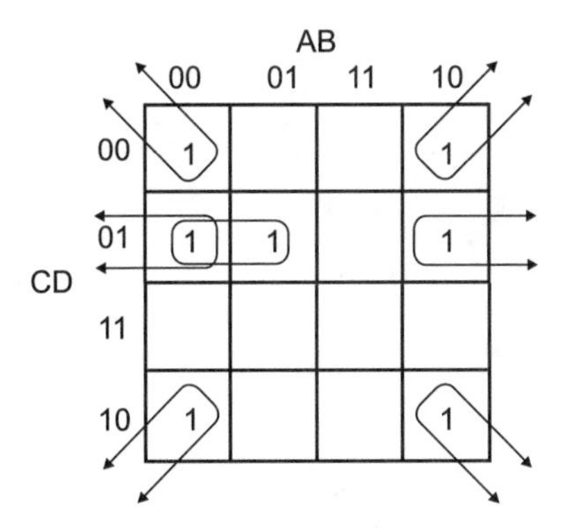


Different grouping gives a different answer.
Different, but equally good.

There are only two common mistakes made with Karnaugh maps. Remember:

1 The numbering is GRAY CODE.
2 The map is continuous in both directions.

## Working straight from truth tables

In the previous example we started with a Boolean expression, but for anyone with a morbid fear of Boolean it may be easier to skip this stage altogether by drawing up the truth table and working directly from that.

To do this, we just ignore any line of the truth table where the result is '0' and plot the lines giving a '1' result, then carry on exactly the same as before. Here is an example.

**Example**

Use a Karnaugh map to simplify the truth table.

Start by picking out the lines in which Q = 1 (lines 3, 4, 7 below).

| A | B | C | Q | |
|---|---|---|---|---|
| 0 | 0 | 0 | 0 | |
| 0 | 0 | 1 | 0 | |
| 0 | 1 | 0 | 1 | ← |
| 0 | 1 | 1 | 1 | ← |
| 1 | 0 | 0 | 0 | |
| 1 | 0 | 1 | 0 | |
| 1 | 1 | 0 | 1 | ← |
| 1 | 1 | 1 | 0 | |

Now draw the Karnaugh map and find the squares referred to by these lines. In our example there will be three squares to find. The first square has A = 0, B = 1 and C = 0, so we locate this on the map and then do the same for the other two lines.

Isn't this just the same as finding the square $\overline{A}B\overline{C}$ on the map? Yes, of course it is, but remember that this method is designed for people who would panic if they saw even a hint of Boolean.

Having located each square on the Karnaugh map we just group them together then write out the simplification. OK, so we finally have to write it as a Boolean term.

**Figure 9.25**
From truth table to Karnaugh

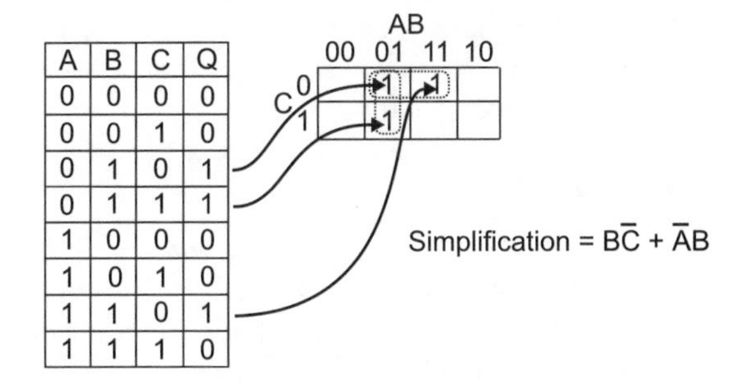


| A | B | C | Q |
|---|---|---|---|
| 0 | 0 | 0 | 0 |
| 0 | 0 | 1 | 0 |
| 0 | 1 | 0 | 1 |
| 0 | 1 | 1 | 1 |
| 1 | 0 | 0 | 0 |
| 1 | 0 | 1 | 0 |
| 1 | 1 | 0 | 1 |
| 1 | 1 | 1 | 0 |

The solution is shown on Figure 9.25.

**Example**

Simplify $\overline{A}\,\overline{B}\,\overline{C}\,\overline{D} + A\overline{B}\,\overline{C}\,\overline{D} + ABCD$.

The Karnaugh map and the answer is shown in Figure 9.26.

**Figure 9.26**
A single square cannot be simplified

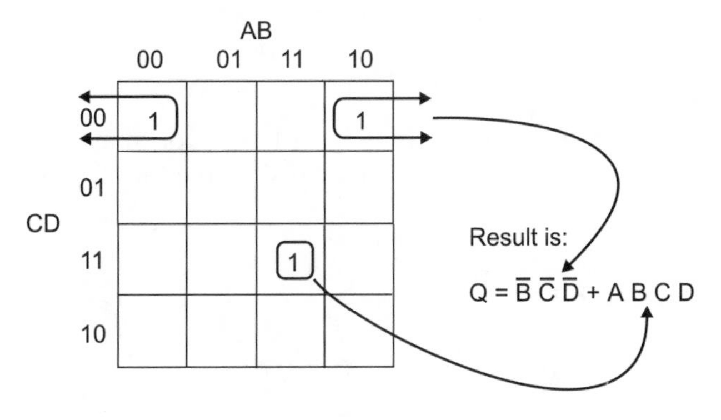


**Example**

Use a Karnaugh map to simplify the expression below.

$$A\overline{C} + ACD + ABCD + A\overline{B}\,\overline{C}\,\overline{D} + AD$$

The answer is shown in Figure 9.27.

**Figure 9.27**
Six squares are best split into two fours

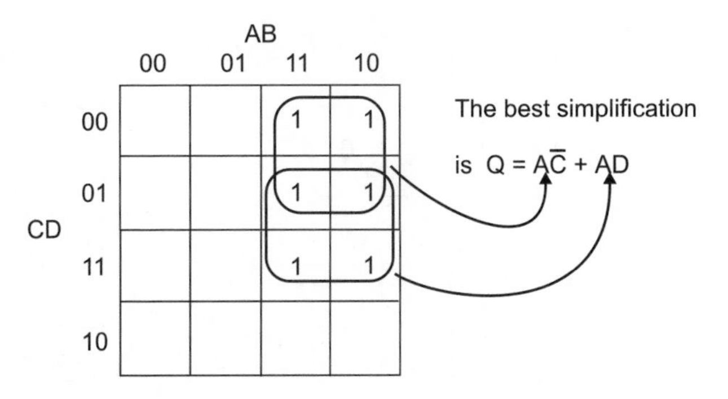

Remember that we can use squares more than once. This allows two blocks of four to be made rather than a four and a two, which would result in a worse simplification.

## 'Don't care' terms

When designing a logic circuit we sometimes meet input combinations that either could not occur or, if they did, the result would not matter. For example, in the logic diagram in Figure 9.28 the input B would have no effect whatever its value, since the zero input at A will be keeping the circuit switched off.

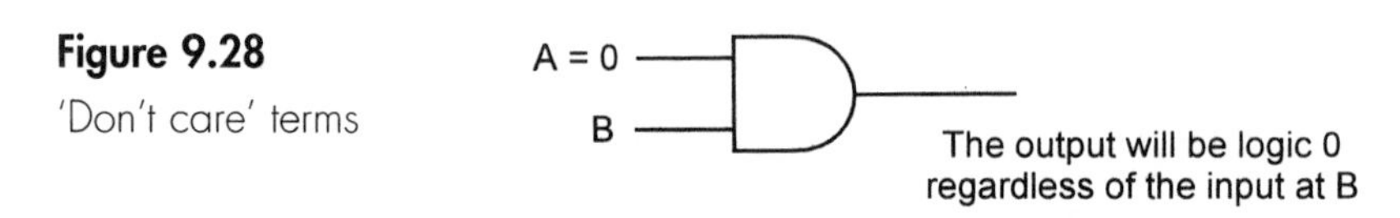


**Figure 9.28**
'Don't care' terms

These 'don't care' states are entered as 'X' rather than 1 or 0, and will prove to be really valuable when simplifying by using a Karnaugh map.

**Example**

Have a look at this truth table. You may recognize this one – it is similar to the one that we simplified in Figure 9.25.

| A | B | C | Q |
|---|---|---|---|
| 0 | 0 | 0 | 0 |
| 0 | 0 | 1 | 0 |
| 0 | 1 | 0 | 1 |
| 0 | 1 | 1 | 1 |
| 1 | 0 | 0 | 0 |
| 1 | 0 | 1 | 0 |
| 1 | 1 | 0 | 1 |
| 1 | 1 | 1 | X |

In this case, however, we have added a 'don't care' state on the last line.

This means that the states

| A | B | C | Q |
|---|---|---|---|
| 1 | 1 | 1 | 0 |

and

| A | B | C | Q |
|---|---|---|---|
| 1 | 1 | 1 | 1 |

are both acceptable.

In the circuit design it probably means that either the A = B = C = 1 state will never occur or, if it does occur, the response of the circuit is of no significance – perhaps the output is disconnected.

On the Karnaugh map, the 'don't care' states are shown as X as in Figure 9.29.

**Figure 9.29**
Adding a 'don't care' square

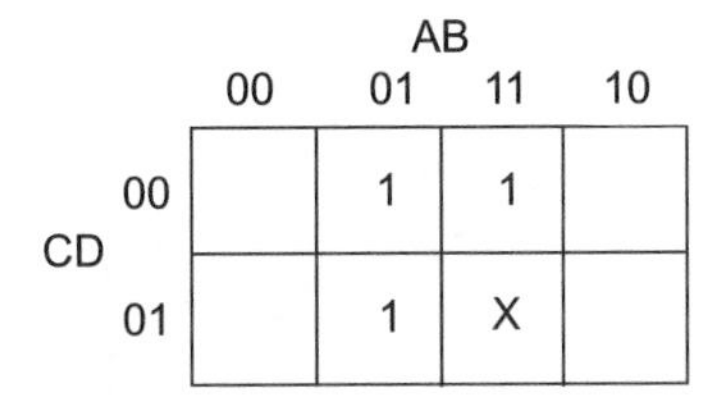


## The choice of the 'don't care' value

When grouping the terms, the 'don't care' term can be taken as a 1 or as a 0 – whichever will give the best grouping.

If it were taken to be a 0 then the overall result would be A = $\overline{B}\,\overline{C} + \overline{A}B$ as in Figure 9.30.

**Figure 9.30**
Assuming the 'don't care' = 0

| C \ AB | 00 | 01 | 11 | 10 |
|---|---|---|---|---|
| 0 | | 1 | 1 | |
| 1 | | 1 | 0 | |

Simplification = $B\overline{C} + \overline{A}B$

If, instead, we took it as a 1 state, then the result would be the much simpler Q = B as in Figure 9.31. That's really good.

**Figure 9.31**

'Don't care' = 1 is a much better choice

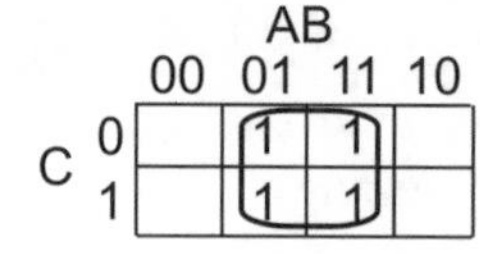


Simplification = B

Remember, when preparing truth table for design work, it is of great benefit if we can spot any 'don't care' conditions and make use of them to help in the simplifying stage.

## Simplification summary

We have three different methods of simplifying a logic diagram:

1 Write out the Boolean equivalent then use the rules of Boolean algebra.
2 Write out the Boolean equivalent then use a Karnaugh map.
3 Draw up the truth table using 'don't care' terms if possible, then use a Karnaugh map.

### Quiz time 9

In each case, choose the best option.

**1 Does a Mercator map of the world, as in Figure 9.12, wrap the same as a 4-input Karnaugh map?**

(a) yes.
(b) no.

**2 Karnaugh maps:**

(a) are always faster than Boolean algebra for simplification.
(b) are generally more certain to find the best solution to a simplification.
(c) can consist of 2, 4 or 6 squares.
(d) cannot be used to simplify Boolean expressions that contain only three terms.

**3 Which of these binary numbers are in Gray code order?**

(a) 0000, 0001, 0010, 0011
(b) 1000, 1001, 1011, 1111
(c) 010, 110, 011, 010
(d) 1000, 1001, 1011, 1010

**4 In Figure 9.32, which of these squares could NOT be paired?**

(a) 4 and 13.
(b) 5 and 8.
(c) 1 and 13.
(d) 3 and 15.

**Figure 9.32**

| CD \ AB | 00 | 01 | 11 | 10 |
|---|---|---|---|---|
| 00 | 1 | 2 | 3 | 4 |
| 01 | 5 | 6 | 7 | 8 |
| 11 | 9 | 10 | 11 | 12 |
| 10 | 13 | 14 | 15 | 16 |

**5 On a Karnaugh map a 'don't care' term indicates a term that:**

(a) is too complicated to bother simplifying.
(b) doesn't exist in the logic diagram.
(c) can have the value of 1 or 0.
(d) is an input to a gate rather than an output.

# 10

# Real gates and their families

The first step in digital repair or design is to gather information about the integrated circuits.

This chapter will consider how gates perform and the various gates that are available.

## How does a gate perform?

Everything that we need to know is provided in data sheets that are freely available from manufacturers, suppliers or in many electronic catalogues. A data sheet is not the most exciting reading available but, like a telephone directory, it can be very useful when you need it.

We will choose a gate and see what we can find out about it. The gate we will use is a 2-input NAND gate. Even this is not so simple since there are so many to choose from, but I have opted for the one with the type number SN74HC00N. We will investigate the numbering system in the second half of this chapter, but for the moment it is enough to say that this is today's most popular 2-input NAND gate for new designs. The official description is a 'quadruple 2-input NAND gate'. The 'quadruple' bit just means that a single IC will contain four 2-input NAND gates, so we buy them four at a time. If we only want to use one gate, we can simply ignore the others.

We will now look at the main items of information contained in the data sheets.

## Description

This simply says what type of gate it is and states the outcome in Boolean algebra, $Y = \overline{A.B}$. It also states the truth table (or function table) and gives the pin-out diagram.

The truth table in Figure 10.1 is shown in a slightly different form to the ones we used in Chapter 5. It uses H (high) and L (low) instead of 1 and 0. By using X for a 'don't care' value the truth table is reduced slightly. Many books and data sheets use H and L instead of 1 and 0.

**Figure 10.1**
An alternative form of a truth table

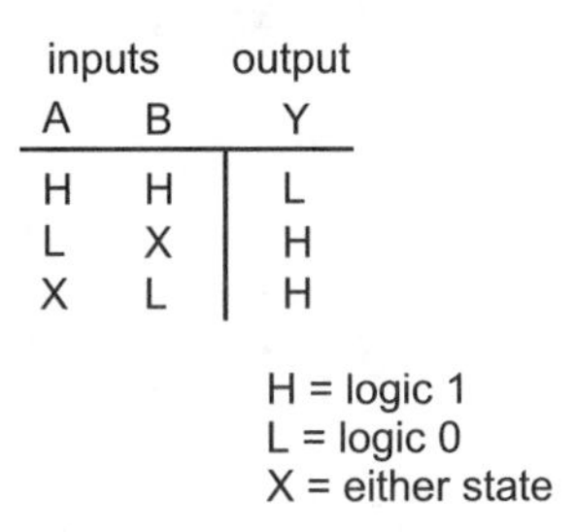
2-input NAND truth table

| inputs A | inputs B | output Y |
|---|---|---|
| H | H | L |
| L | X | H |
| X | L | H |

H = logic 1
L = logic 0
X = either state

Notice that the supply voltage, called $V_{CC}$, is on pin 14 and the ground or 0 V connection is on pin 7. This is a fairly standard layout for the supplies and the four NAND gates are shown with A and B as their inputs and Y as their outputs (Figure 10.2).

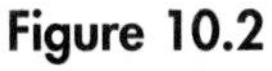
**Figure 10.2**
The pin-out diagram of an SN74HC00N

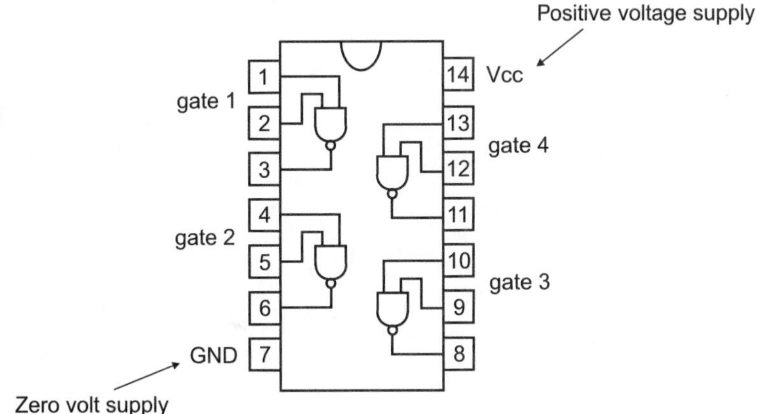


## Supply voltage ($V_{CC}$)

This is the supply voltage that should be used if the performance and reliability is to be guaranteed. In our case, the nominal supply voltage is 5 V with a minimum value of 2 V and a maximum value of 6 V.

## High-level input voltage ($V_{IH}$)

This is the lowest voltage that the gate will accept as a logic 1 (or logic 'high') level. It sometimes varies according to the supply voltage. In this chip the minimum input level changes from 1.5 V when $V_{CC}$ = 2 V up to 4.2 V when $V_{CC}$ = 6 V.

Incidentally, the symbols like $V_{IH}$ can usually be interpreted by taking the terms in order. This one starts with V, so we are talking about a voltage. The $_I$ stands for input and the $_H$ stands for high (or logic 1). Most terms can be resolved in this way.

## Low-level input voltage ($V_{IL}$)

This is the highest voltage that will still be accepted as a logic 0 (logic L). Again it depends on the supply voltage, and in our example it varies from the range 0–0.5 V and from 0–1.8 V as the supply changes from its minimum to maximum values.

## Noise margin

Between the voltage corresponding to a level 0 and that of level 1 lies a no-man's-land which we try not to use. If we apply such a voltage we do not know whether the gate will accept it as a level 0 or as a level 1, and the output will be quite unpredictable.

Noise is the name given to a random variation in a voltage level. It may be caused by external interference, such as from nearby switches or electric motors, or atmospheric noise. It can equally well be caused by random electron movements due to heat within the semiconductor from which the gates are made. The result is shown in Figure 10.3.

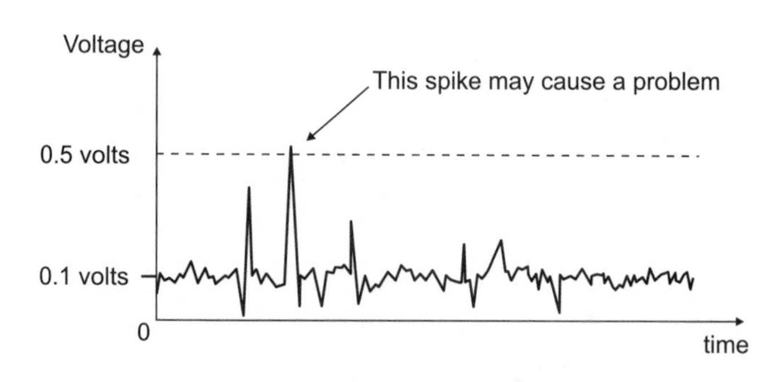


**Figure 10.3**
A 'noisy' voltage

If we connect two gates together, the output of the first must be acceptable as an input to the second gate. In the situation shown in Figure 10.4, the first gate can provide an output logic 0 which is anywhere between 0 V and 0.1 V. The second gate requires a logic 0 input somewhere between 0 V and 0.5 V.

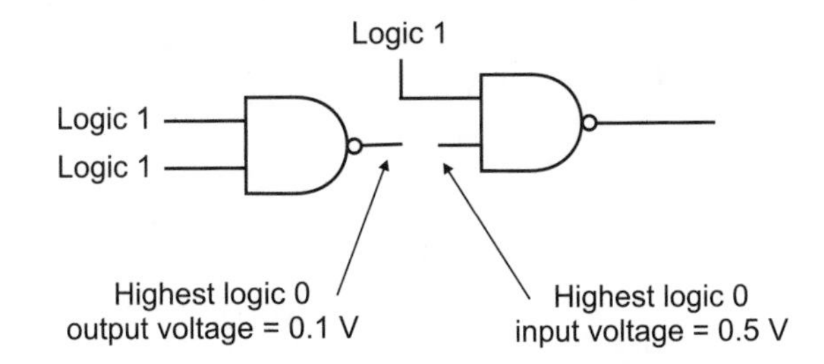


**Figure 10.4**
The worst-case scenario

In the worst case, we could have the first gate producing its highest acceptable level of 0.1 V, leaving a gap of only 0.4 V. This gap is called the noise margin, and represents the highest value of noise that the system can handle.

At logic 1 we have a similar situation since the first gate must provide at least 1.9 V and the second gate must accept at least 1.5 V, leaving another noise margin of 0.4 V. In this example both noise margins happen to be the same value, but this does not always happen. When the noise margins have different values, the data lists either both of them or just the lower of the two. This is summarized in Figure 10.5.

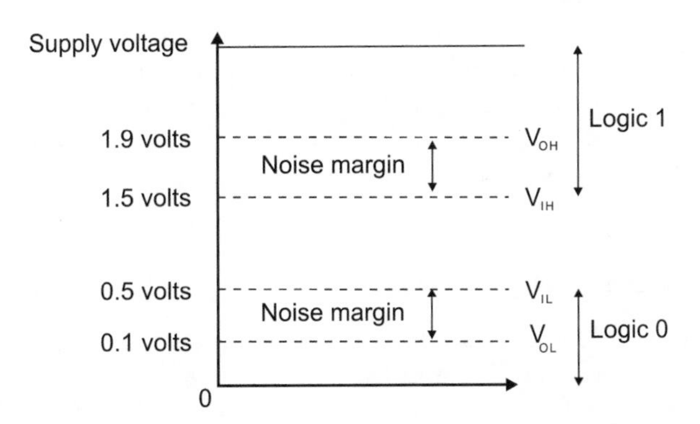


**Figure 10.5**
Typical noise margins

## Supply current ($I_{cc}$)

This is the current that is needed to work the internal circuit. It is the current flowing into the $V_{CC}$ pin. In our example chip, this supply current is 20 μA.

This appears to be a very simple parameter to measure and so it is . . . but. The 'but' here is to remind us that it is a competitive market and, like statistics and politics, the statements may be absolutely true but what is NOT said is equally important. Just as with politicians, we are sometimes fed with selective truths and then left to draw our own wrong conclusions. Even if all suppliers have quoted the same figures, this still may not coincide with what happens when we use the chip to build a real circuit.

Before accepting the figure we first have to ask four questions:

1 Is the current quoted per gate or per package? In the SN74HC00 there are four NAND gates, so we could 'reduce' our current by saying that the current is only 5 µA and the small print could state 'per gate'.
2 What are the logic states being used at the time of measurement? Some gates pass different currents under different logic conditions.
3 At what frequency is the chip switching? Or is at rest, called the quiescent state? Some designs of gate are unaffected by switching rates, but in other designs this is a crucial factor. In one example the current varies from almost zero in the quiescent state to one of the highest values of current at high frequencies.
4 What supply voltage ($V_{CC}$) was used for the test?

## Input current logic high ($I_{IH}$) and logic low ($I_{IL}$)

These are the currents measured at the gate inputs – not the supply currents.

In the earlier designs with the 'LS' type numbers, the $I_{IH}$ current has a value of about 20 µA and the $I_{IL}$ current has a value of about 0.4 mA. In more recent designs the input currents have fallen to values of less than 20 µA and in the later ones, after the 74HC family, the current will not exceed 5 µA. In the case of the $I_{IL}$, the earlier families were less than 100 µA and the later ones are again less than 5 µA. So if we ignore the 74 and 74LS families, we can forget about input currents.

## Output current ($I_O$)

This is a question of how much current the gate can supply to another circuit. Perhaps it is controlling an output display or other gates. The value of this current may differ according to the logic state of the output. Some gates can provide a higher current when the output is at logic 0 than when it is at logic 1, whereas in many of the more modern gates the current levels are the same in each case. See Chapter 11 for more on this.

## Rise time ($t_R$), fall time ($t_F$)

These terms are a measurement of how fast a waveform can increase or decrease in value. It may be measured at the input to a gate or at the output. This is a general electronic measurement rather than something dreamt up just for logic gates. The measurement points are always taken at the 90% and 10% values of the waveform, as shown in Figure 10.6.

**Figure 10.6**
Speed of voltage change

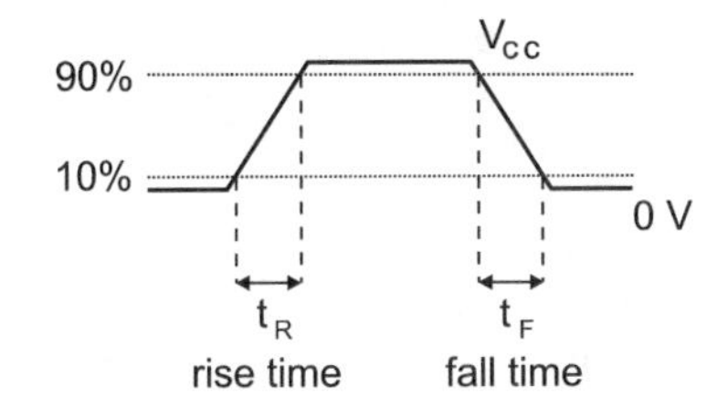


Associated with this is the slew rate of a waveform or a device. This is based on the same information but is converted to volts per second, so if a voltage changes by 2 V in 50 ns it will have a slew rate of

$$\frac{\text{volts}}{\text{time}} = \frac{2.0}{50 \times 10^{-9}} = 40 \times 10^{6}$$

or 40 MV/s or 0.4 V/ns. If the rise and fall times are different, check to see which has been used to calculate the slew rate.

## Propagation delay ($t_{PD}$)

When we switch a gate by changing the input voltage there is a very slight delay as the internal circuitry responds to the new conditions. Typically this will take a few nanoseconds (ns, $10^{-9}$ seconds).

Once again the conditions of the test are significant, and propagation delay is measured when the output is at the 50% value when changing from low to high as well as from high to low. In some cases this results in the same value, but not always.

The symbols used are $t_{PLH}$ (time for propagation from low to high) and its partner $t_{PHL}$ (time for propagation from high to low). An example is shown in Figure 10.7. The measurements are taken at the same points whether or not the gate inverts the signal.

If $t_{PLH}$ and $t_{PHL}$ result in the same values, the single figure may just be listed as $t_{PD}$ (PD = propagation delay).

In some data sheets the propagation delay is simply listed as 'speed'.

**Figure 10.7**
Propagation delay

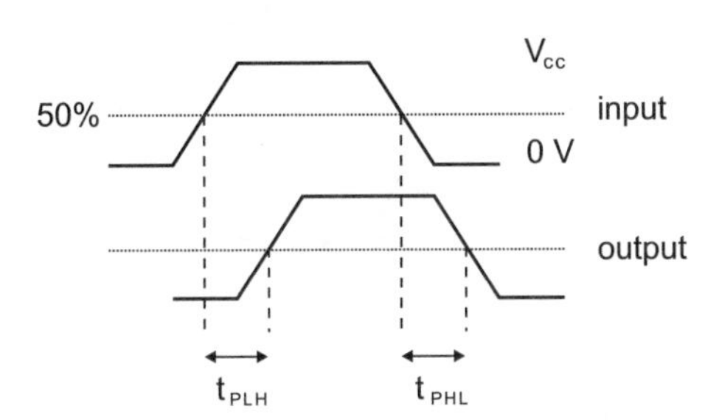

## Absolute maximum ratings

The data sheets usually describe the normal operating parameters, that is, the limits within which the IC is guaranteed to perform to the whole of its specification. In addition to this, they sometimes list the 'absolute maximum ratings', which are the limits beyond which the device is likely to be destroyed.

In the case of our SN74HC00N, the recommended $V_{CC}$ is given as 2–6 V but the absolute maximum rating for the $V_{CC}$ is –0.5–7 V. This doesn't mean that we can operate the chip between 6 V and 7 V without a great deal of thought. For example, increasing $V_{CC}$ will increase the power dissipated inside the chip and hence its temperature. It may all end in tears, and the chip supplier will be totally uninterested in any grumbles that you may have.

## Power dissipation ($P_{diss}$)

This is the measure of how much power can be dissipated. Be careful to check whether the stated figure is per gate or per chip, and what the ambient temperature was at the time of measurement.

This completes our roundup of the main parameters and gives us a reasonable basis for comparing different ICs. There are other, less important, items of data that will be supplied with an IC, but life is going to get incredibly boring if we do any more of this so I think it is a good time for a full stop.

## What gate is this?

Each individual design of integrated circuit is given its type number. There may be several different numbers printed on the integrated circuit – some indicate the assembly plant, the supplier or the manufacturer's batch number – but one of them will be in the general format shown in Figure 10.8, which is our magic number. This number provides us with all the information that we need to discover absolutely everything about the chip.

**Figure 10.8**
The starting point

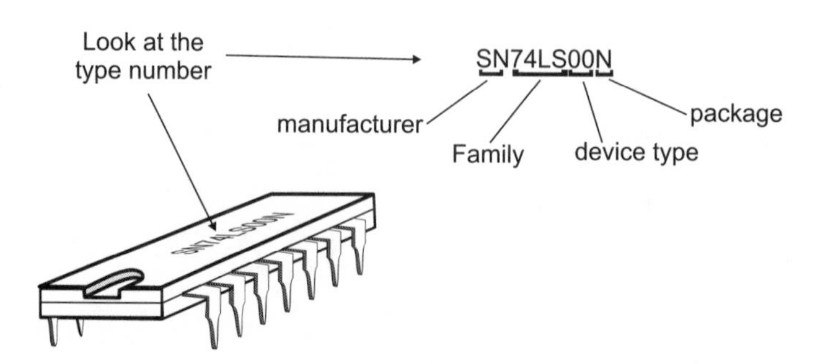

A typical type number of SN74LS02N is shown, and is dissected below.

## The letters at the start

There are many manufacturers, and they usually add between one and four letters or a number followed by more letters to indicate the manufacturer. Sometimes two manufacturers share a code. Thus both Texas and Motorola have used the letters SN. Most manufacturers use several different code letters for different products. The first letter or letters are therefore not of great interest to us since they all comply with the same basic specification.

## The family number

Some logic designs have been more successful than others and have become accepted as 'industry standard'. There is no point in designing a really good AND gate unless you also have ORs, NANDs etc. to form a complete set, called a 'family' or 'series'.

As with most technology, the process of continuous development causes new families to be born and old ones to wither away. The withering process can be a slow one, and some families in everyday use are well over 20 years old and are still being sold.

In the chip SN74LS00N it is the 74LS that tells us the basic design and is the family name.

A little later on we will look at the current state of affairs in three groups: the old folk, those in the prime of life and, finally, the youngsters.

## Device type

After the family number there is another group of numbers.

These numbers tell us what actual device we have. For example, 00 means a NAND gate in the 74 series, and 86 is an XOR gate. In the 4000 series the numbers are different, and these two types of gate are numbered 11 and 70 respectively.

It is worth mentioning at this point that all manufacturers employ the same type numbers, so a chip called 7400 is a NAND gate from any supplier and is completely compatible.

## Package

The actual circuit is built on a small piece of semiconductor material a few millimetres square. To handle it and make the necessary connections we have to enclose it in a holder of some kind.

One of the most popular is the plastic or ceramic package shown in Figure 10.8. This package is called a d-i-l (dual-in-line) package, and is either plugged into a base or pushed through holes punched into the printed circuit board and soldered on the underside of the board. Most chips now have a 'surface mount' option. These are very much the same in appearance except that the pins are shorter and do not penetrate the printed circuit board but are soldered on the top surface of the board (see also Chapter 20).

A final letter indicates the type of package. The available types are common to all manufacturers, but the letters used to signify them are not.

The package letter sometimes also includes another letter added to indicate the temperature range of the device.

## The development of logic

In this section we are going to limit the discussion to a few of the more common families. They are the ones that we are likely to meet and a few of the new types that may, perhaps, be the winners in the next decade.

## The old folk

These are displayed in Figure 10.9. They were generally designed in the 1970s. Some have inevitably been overtaken and have faded away, but some are still very much alive.

**Figure 10.9**
The old folk

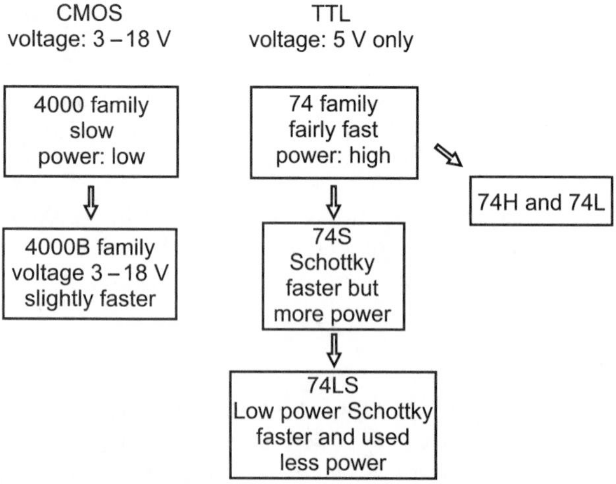

CMOS
voltage: 3 – 18 V
4000 family
slow
power: low
4000B family
voltage 3 – 18 V
slightly faster

TTL
voltage: 5 V only
74 family
fairly fast
power: high
74H and 74L
74S
Schottky
faster but
more power
74LS
Low power Schottky
faster and used
less power

## The 74 TTL family

The most common series ever is the 74 series. This accounts for the large majority of all digital integrated circuits in use and includes all the common gates. We are all used to the speed of technological change, so it is interesting to know that the 74 series was first produced in the early part of the 1970s and is still listed in the current catalogues. There are generally two grades of device in this series, called 'commercial' and 'military'. The principle difference is in temperature tolerance and reliability (and price!). The usual 74 series can only operate down to 0°C and up to +70°C. The lower limit would be a decided disadvantage in military campaigns in arctic regions or even in temperate zones during the winter months, and so the military family was extended to operate from –55°C to +125°C.

To distinguish between them, the family number 74 was given to the commercial version and the number 54 was applied to the military version. If you are involved in repairing old digital circuits, you may come across the obsolete 64 series which, as the number suggests, lies between the commercial and military families. This was called the 'industrial' family, and had an intermediate temperature range. These three families were all plug-in replaceable, which means that all the pins are used for the same jobs and require no rewiring of the circuits.

The original 74 series was a design based on a transistor circuit, and was referred to as a TTL (transistor transistor logic) device. TTL logic gates operated on a supply voltage of 5 V; this became the standard voltage for digital circuitry and so it has remained for 20 years. It also established that in the 74 series, logic 1 was to be any voltage over 2.0 V and logic zero anything less than 0.8 V. The voltage ranges of the original 74 series have set the trend for many of the more recent arrivals.

## CMOS 4000

The other family that has become a 'standard' is the CMOS 4000 series – usually just called the 4000 series. The 'CMOS', incidentally, refers to 'complementary metal oxide semiconductor', and refers to the design features.

The 74 and the 4000 series swept away all the previous families and virtually gave us a fresh start. The earlier series are of no interest to anyone except a museum curator.

The CMOS 4000 family and the improved version, the 4000B, were introduced shortly after the 74 series. Their main selling points were the supply voltage, which could range from 3–18 V, and that they required only a tiny operating current. These made them ideal for direct connection to a standard 9-V battery. The supply current is actually dependent on the operating frequency being nearly zero at very low speeds and increasing steadily as the frequency increases, but is still low compared with the 74 series, which stays at a constant value regardless of operating frequency.

The drawbacks of the 4000B family were its very low operating speed, about 5 MHz, and that its logic levels when operated at 5 V were different to the 74 series. Logic 0 ranged from 0–1.5 V and logic 1 was from 3.5–5 V. They also had different pin-outs. This meant that they could not easily be combined with 74 TTL gates in a circuit.

A few similar variants, 4100, 4500 and 4700, were also produced. These families are often grouped as the 4XXX family, the X being commonly used in all digital documentation to indicate optional numbers, values or logic states.

## More speed

At this stage we started to look for improvements.

Ideally what we were after was a family that delivered high speed, used very little power, and was compatible with previous designs.

Our first attack was on the question of speed.

## 74H and 74L

The first attempt at improvement was the 74H family. To make it switch faster, we would have to change the internal voltages faster. To do this, we could decrease the value of internal resistors. This had the disadvantage of increasing the power consumption and the 74H (H for high speed) did not survive. We also tried to reduce the power consumption by increasing the resistor values, and the 74L was born. It soon became apparent that the L version was too slow, and this family also died away.

## 74S

Having fiddled with the resistors without much benefit, we turned our attention to the transistors. A new design of transistor called the Schottky transistor, or more correctly the Schottky-clamped transistor, was developed which had the ability to switch between on and off states at very high speeds by preventing saturation of the transistors. This, together with low resistor values, produced high speeds, but unfortunately it still consumed a lot of power. The 74S family (S = Schottky) has survived (just). Its main function was to set the course for the all-time winner, the 74LS family.

## 74LS

This family reduced the power of the original 74 series by a factor of five down to 2 mW per gate and also improved on the speed, allowing it to switch up to 25 MHz. This one made the basic 74 family virtually obsolete. After its launch in 1976 this family became the most popular, and it is still in widespread use today.

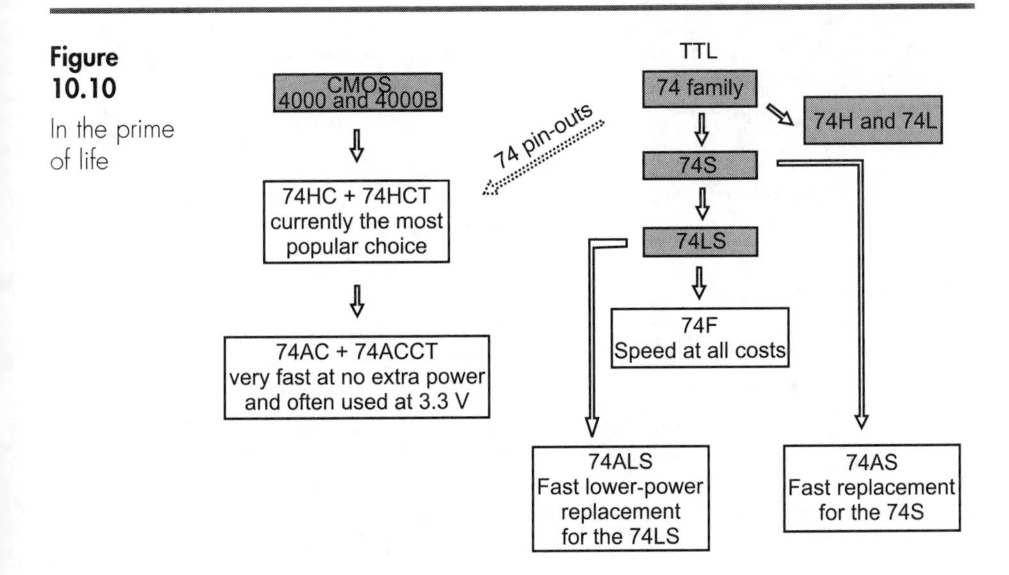


**Figure 10.10** In the prime of life

## In the prime of life

These are today's (1999) best sellers and will become tomorrow's old folk. They are collected together in Figure 10.10.

### 74F

This is an improved Schottky design (F = fast), in which we went for speed and achieved 100 MHz at the expense of the power dissipation increasing to over 5 mW/gate. This was quickly followed by two rivals, the 74AS and the 74ALS.

### 74AS and 74ALS

The first was the 74AS (advanced Schottky), which was slightly faster but used 50% more power and was intended to be the replacement for the 74S. The other one was the 74ALS (advanced low-power Schottky), which was designed to replace the 74LS and was completely compatible, but the speed had increased from 25 MHz to 34 MHz and the power consumed had gone down from 2 mW/gate to 1.3 mW/gate.

### 74HC and 74HCT

The HC in the title indicates that these are high-speed CMOS devices, and the HCT means high-speed CMOS TTL-compatible. These CMOS families departed from the 4000 series in their choice of supply voltage. The wide voltage range of the earlier CMOS designs was abandoned and reduced to 2–6 V, thus allowing them to operate at 5 V just like the others in the 74 series. The significance of the TTL-compatible part is that the voltage levels of the supply and the logic levels were made to coincide with the original 74 series. Both versions were made compatible with regard to the pin-out diagram so that the HC family can be used without redesigning any circuits. At the moment (1999) the HC family is the most popular choice for all new designs.

Both HC versions are slightly faster than the 74LS family and consume only about one-quarter of the power.

#### 74AC and 74ACT

These advanced CMOS devices were a significant improvement over the HC family, increasing their operating frequency up to 125 MHz without any increase in the power consumption. There was another change occurring at that time – there was new pressure to reduce the operating voltages of integrated circuits. We had found that a decrease in the supply voltage could have a knock-on effect of increasing the possible operating speeds. This has now been reflected in modern microprocessors, which have also reduced their supply voltage to 3.3 V in the search for more speed. (There is more about this in the companion book *Introduction to Microprocessors*.) The 74ACT family is the TTL logic level version with the other characteristic unchanged.

### The youngsters

So far we have been looking at history and how things have developed and what has proved popular – this has been easy. We now have to take a guess at the future – time to get things hopelessly wrong. However, we will look at four of the likely contenders from the up and coming generation.

The general trend is towards more speed (of course!), more output current and less power consumption, and a 3.3 V operating supply. These are shown in Figure 10.11. The 74 pin-outs remain, but the 3.3 V designs are clearly the future.

#### 74LVC and 74ALVC

These CMOS devices are approaching the edge of the universe as we know it. They are designed for operation at 3.3 V supply voltage, hence the name LVC (low-voltage CMOS) and ALVC (advanced low-voltage CMOS). There are few devices available as yet. The A version has a propagation delay of under 3 ns and offers 100 MHz performance with moderately high output drive currents.

#### 74AHC and 74AHCT

These are produced jointly by Texas Instruments and Philips. The AHC stands for advanced high-speed CMOS, and the T is obviously the version with modified logic levels to make it TTL-compatible. Of the two, the 74AHC is the better chip to use if it is not to be mixed with

**Figure 10.11**

The youngsters on the edge of the universe

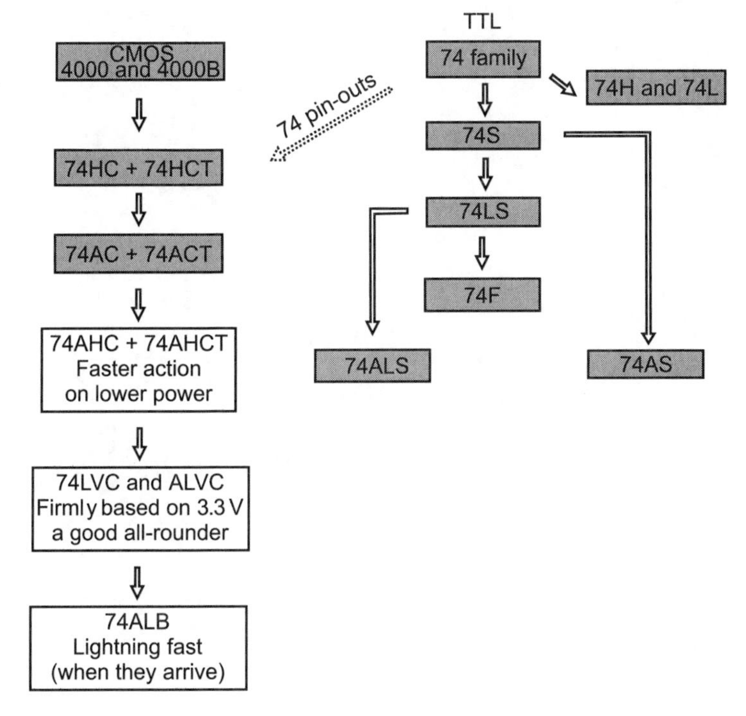


other TTL chips. This advice applies in all families where a TTL version is available. This family offers about half the power consumption and half the propagation time of the 74LS family whilst running at a frequency of 150 MHz or thereabouts.

And finally . . .

### 74ALB

There are few devices yet designed to add to this family, but they are very promising although we need a little hindsight to see how successful they will be. At the moment they are certainly state of the art. The ABT (advanced low-voltage BiCMOS) family offer propagation delays of under 2 ns! They operate on 3.3 V, and the present devices are designed to work with high-speed cache or main memory in computers needed for multimedia and high-speed graphics applications.

## Performance summary

It must be emphasized that the figures quoted in Table 10.1, and other sources, are only guides to allow comparisons to be made. The actual figures vary according to the conditions under which they were measured, and which actual device was taken as representative of the

**Table 10.1** Performance summary of gate families

| | *4000* | *74LS* | *74F* | *74AS* | *74ALS* | *74HC* | *74AC* | *74LVC* | *74AHC* |
|---|---|---|---|---|---|---|---|---|---|
| V range (V) | 3–18 | 4.75–5.25 | 4.5–5.5 | 4.5–5.5 | 4.5–5.5 | 2–6 | 2–6 | 2–3.6 | 2–5.5 |
| I output high (mA) | 0.4 | 0.4 | 1 | 2 | 0.4 | 4 | 24 | 24 | 8 |
| I output low (mA) | 1 | 16 | 20 | 20 | 8 | 4 | 24 | 24 | 8 |
| I at f = 0 Hz (μA) | 0.06 | 3 | 6.5 | 10 | 1.9 | 0.02 | 0.02 | 0.01 | 0.02 |
| Tpd max (nS) | 100 | 15 | 6 | 4.5 | 11 | 20 | 8.5 | 4.9 | 8.5 |
| i/p levels (V) | CMOS | TTL | TTL | TTL | TTL | CMOS | CMOS | TTL* | CMOS |
| o/p levels (V) | CMOS | TTL | TTL | TTL | TTL | CMOS | CMOS | TTL* | CMOS |
| f (max) (MHz) | 5 | 40 | 100 | 25 | 34 | 40 | 100 | 100 | 125 |

*when using a supply voltage of 3.3 V

whole family. In addition to these approximations, each gate is subject to variations caused by normal manufacturing tolerances.

Before you use a particular chip, make sure you check the figures against the data sheet supplied by the manufacturer.

The families with TTL versions such as the 74ACT, 74AHCT etc. generally differ only in the recommended power supply voltages. The typical CMOS values of 2–6 V are modified to the TTL values of 4.5–5.5 V.

Remember that the CMOS chips increase their supply current and hence power dissipation as the frequency increases, whereas TTL devices remain at the same values, near enough, at all frequencies. The supply currents quoted in data sheets often refer to the quiescent state (0 Hz) and are therefore very kind to CMOS devices. At high frequencies, the differences are less significant.

## Quiz time 10

In each case, choose the best option.

### 1 In the type number TC74LS08N:

(a) the letters TC mean that the construction is low propagation time CMOS.
(b) the numbers 08 indicate the family.
(c) the letters LS mean low speed.
(d) the letter N indicates that the package type is dual-in-line.

### 2 The last letter in a chip number indicates its:

(a) package type.
(b) date of first manufacture.
(c) family.
(d) place of manufacture.

### 3 In their quiescent state, many chips:

(a) tend to overheat.
(b) consume less power than at higher frequencies.
(c) have a very low propagation delay, which increases at higher frequencies.
(d) of the 74ALS family have floating outputs.

### 4 A TTL gate is likely to have a supply voltage of:

(a) 5 V.
(b) between +2 V and +6 V.
(c) between +3 V and +18 V.
(d) 3.3 V.

**5 Which of these gates is the slowest?**

(a) 74ACT.
(b) 74S.
(c) 4000.
(d) 74LVC.

# 11

# Interfacing

Interfacing is the connection of one circuit to another.

There are only two requirements:

1 It must work.
2 It must not damage either circuit.

The general method is to decide what you want to do and then study the data sheets point by point to see if there is any clash in the requirements.

It sounds easy and sometimes it is, but not always. Sometimes we come across a snag that we have forgotten or never knew about.

There are possibly four important areas that cause problems:

1 How fast?
2 How much current?
3 What voltage?
4 Things we may have overlooked.

## How fast?

### Chip to chip

If we designed a circuit to run at 125 MHz using the 74AHC family, it would be a bit of a disaster to put a 74AS chip in the middle that could only run at 25 MHz. The whole circuit can only run at the speed of the slowest device and this is the electronic equivalent of the popular adage that a chain is only as strong as the weakest link.

## Chip to something else

When we interface a digital circuit to another type of circuit, we may well find that the maximum clock speed is quite irrelevant. Take us, for example; there is nothing quite like us to slow a circuit down. A logic gate is amazingly fast compared with us. If we feel the temperature of a piece of metal and it is too hot, we immediately take our hand off. But how long did this take? For most people the time to think and then respond would be about one-tenth of a second.

Imagine you are a goalkeeper in a premier league football match and you are facing a penalty. The ball is placed on the spot. The ball will enter your goal half a second after the moment it is kicked. Now, if you take a tenth of a second to decide which way to go, this leaves you only the remaining four-tenths of a second to intercept the ball in the top corner. This means that you must average 35 km per hour from a standing start. Millions of 'armchair internationals' groan as you miss such an easy shot. In reality, your only chance is to use body movement or magic to predict what is going to happen. Waiting to respond to the flight of the ball will make you too late.

Imagine that a logic circuit is controlling a coffee machine. We put our money in and it flashes up a message saying 'Select item required' and we have to press a button. At our fastest, we could hit the button in one-tenth of a second. So what can the logic circuit do in the same time? A modern logic gate can respond to a button in about one-hundredth of a microsecond, or 10 000 000 times faster than us.

This means than the logic circuit lives at a speed about 10 000 000 times faster than we do.

Can you imagine how we would feel faced with a creature called a 'hangabout' that runs ten million times slower than us?

It would take over 11 days for it to press the button. After all that effort, it may run off at 1 mm per hour to spend 285 years having a cup of coffee. By way of compensation, it may well live for 800 million years!

Whenever the logic circuit is interfacing with something that is not electronic, it is very likely that the speed of even the slowest gate is going to be more than enough.

## How much current?

## Chip to chip

### Sourcing and sinking

Conventionally current is said to flow from positive to negative, so if the output of a gate is held at logic 1 then current will flow from this positive output through the load and down to the 0 V supply. This current is supplied or sourced by the IC, and is referred to as a source current. However, if the output of a gate is held at logic 0 then current will flow from the positive supply through the load and into the gate. This current is referred to as a sink-current, and the IC is said to be sinking the current. So it is just a matter of which way the current is flowing (Figure 11.1). In some families, particularly the older designs, the sink current is larger in value than the source current. Not only do the newer designs have the same values for both of these currents, but also the value of the current has generally increased.

**Figure 11.1**
Sourcing and sinking

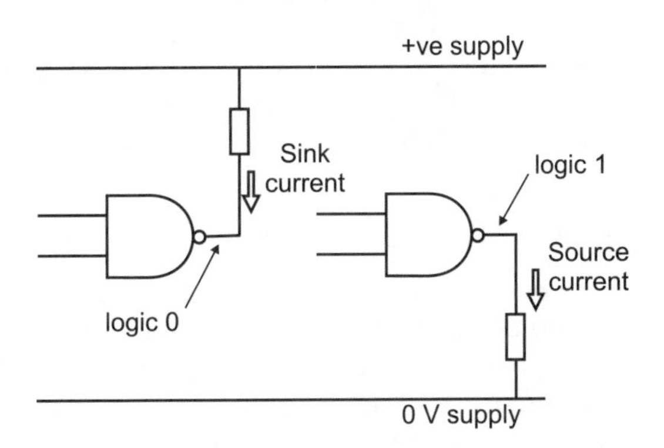


As a convention, the data sheets refer to the sink currents as positive and source currents as negative.

## Fan-out and fan-in

How many gates can be connected to the output from a single gate? The traditional answer to this was a matter of comparing the output current supplied by a gate with the input current needed by each gate connected to it. If the available output currents are, say, 10 times greater than the input currents, then we can reasonably expect to be able to connect 10 gates to the output of one. In this case the driving gate would have a fan-out of ten.

This was true while we were dealing with the early 74 and 74LS series. In the case of the 74LS family the output can supply 10 times the input current, so each 74LS chip can drive 10 other members of the same family. In the 74LS family we can state 'fan-out = 10'. In more modern families the output currents are so much higher than the inputs that the drive current is seldom a problem.

Fan-in is much the same except that we are looking at how many gates can be connected to the input of a gate. This limit is usually imposed by the output capacitance of each gate. As more are added in parallel the total capacitance increases, and this reduces the speed at which the input voltages can change.

**Historical note: unit loads**

In some data sheets we come across the sink and source currents expressed in terms of unit loads rather than milliamps. This was very popular in the early days, but is now obsolete and has been replaced by the actual current values.

When the original 74 series were first developed, the input current in the high state ($I_{IH}$) was 40 µA and the input current in the low state ($I_{IL}$) was 1.6 mA. These two values were considered to be rather special and were given the name 'unit loads'. The high state output current had a value of 20 unit loads (0.8 mA) and a low state output current of 10 unit loads (16 mA). It was intended that, by quoting all gate input and output currents in unit loads, we could see if a design was viable just by adding up all the unit loads connected to the output and hoping that it did not exceed the unit loads available.

## Chip to something else

If a circuit requires an input of 20 mA then it is no good using a CMOS 4000 device that, at best, can only provide a 1 mA output. The other way round is OK. It wouldn't matter if we used a 20 mA gate to provide a current of 1 mA because the 20 mA figure is just the maximum capability of the chip, it doesn't have to pass that value of current.

In many cases, the logic output is used to provide the input to an external current amplifier and that, in turn, drives the final circuit.

## What voltage?

Voltage levels are much the same, but here we may meet two problems. A circuit designed to use the CMOS 4000B family can have a supply voltage of +18 V. If we chose to put a 74LS chip into this circuit it would be one very unhappy chip indeed, since these chips can only work up to +5.25 V. The other voltage problem is a matter of the logic levels. An output at logic level 1 must be accepted as a level 1 input by the next IC. The same applies to logic 0 levels, of course. The logic levels are compared in Figure 11.2.

# Things we may have overlooked

## Switch bounce

We often use a switch to provide an input to a gate circuit. Now, we like to think that the switch just changes the voltage level at the input to the gate, but the contact actually bounces just like a ruler twanged on the edge of a table by generations of schoolchildren. The output voltage changes between logic 1 and 0 maybe 50 times over a millisecond or two. As the switch is closed the input to the gate is

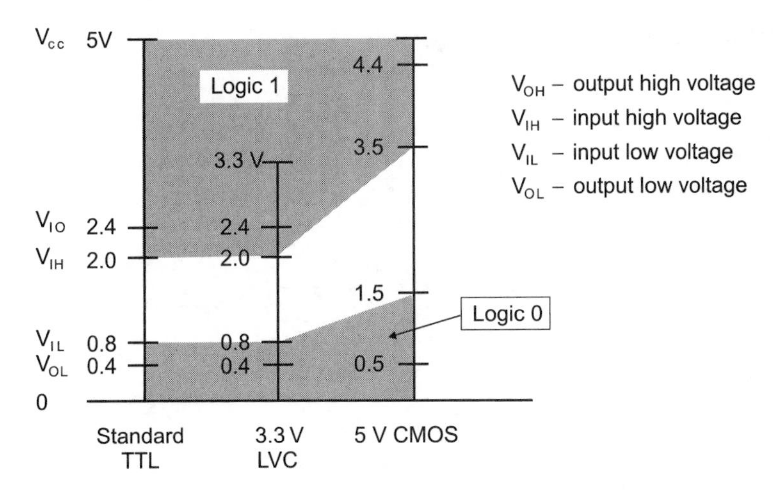


**Figure 11.2** CMOS levels can cause problems

grounded, but the contact actually bounces up far enough to disconnect the 0 V input. At this moment the gate input is returned to the +5 V supply or is left floating, which has much the same effect. The contact now springs back down again to reconnect the 0 V supply only to bounce off again. This continues until it finally settles at 0 V. Only a few bounces are shown in Figure 11.3.

A glance through a supply catalogue will find a variety of bounce-free switches. In the next chapter we will look at a simple circuit for converting any cheap bouncy switch into an expensive bounce-free one.

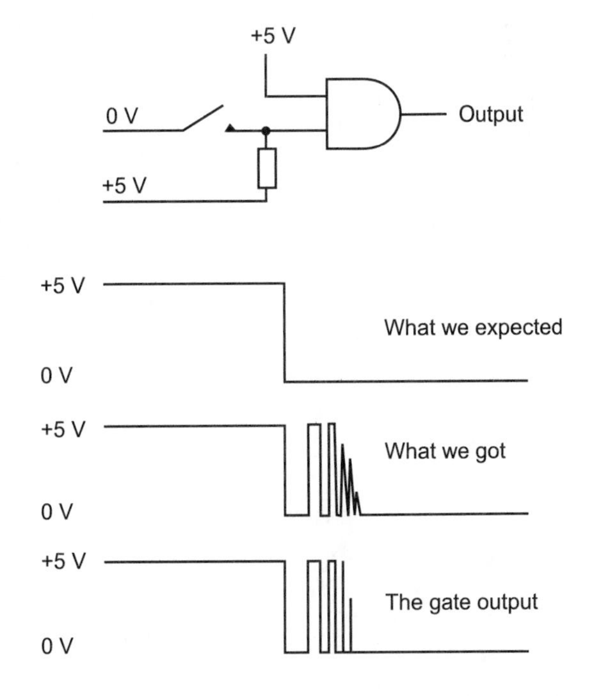


**Figure 11.3** Switch bounce

## Connecting outputs together

Figure 11.4 shows the problem. What if we switched one gate on and the other off? Well, it all depends on the design of the gate.

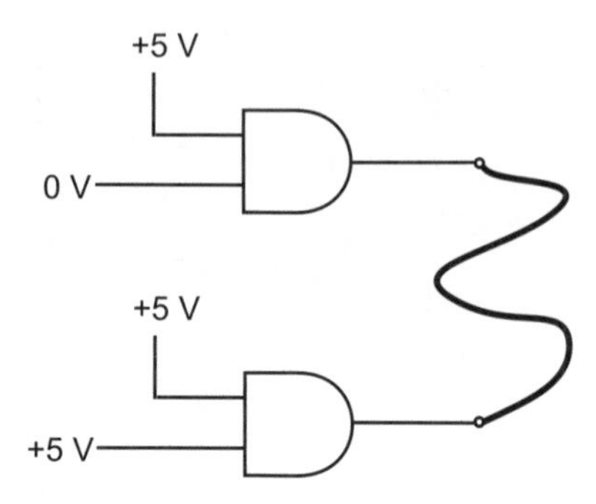


**Figure 11.4**
Smiles or smoke?

### Totem-pole outputs

Most gates have totem-pole output stages within the chip. This is the name given to a gate output design that consists of two transistors connected in series. A transistor is just like an electronic switch that can be either on or off. These two transistor switches are connected across the supply, and the output is taken from the mid-point. The remainder of the circuit is used to switch these transistors as required. One or other is on, but never both. In Figure 11.5 the output is connected to +5 V to give logic 1 output. If we had two gates, one with logic 1 at the output and the other at logic 0, a problem would arise if we connected the outputs together (Figure 11.6). As we can, see there is a short circuit across the supply lines and there would be excessive current flowing through the two transistors until one or other of them overheated and was destroyed.

For nearly all gates, connecting the outputs together is a disaster.

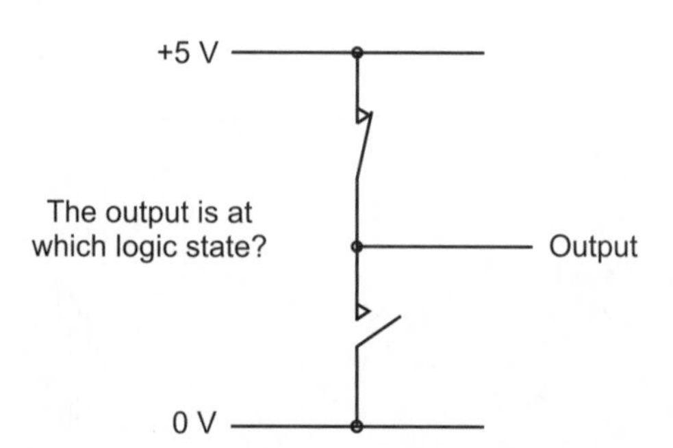


**Figure 11.5**
A totem-pole output

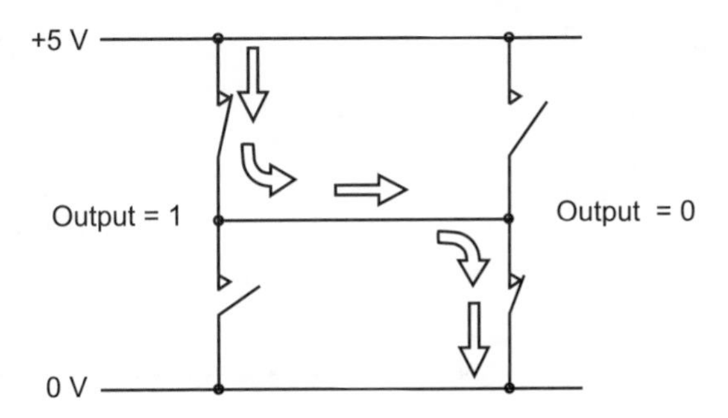


**Figure 11.6**
A totem-pole disaster

## Open collector gates

Open collector gates in TTL or the equivalent open drain gates in CMOS have been modified slightly to allow their outputs to be connected. They will always be clearly signposted in catalogues and data sheets with a separate number. An example of a normal 2-input AND gate is listed as 74LS08, and the open-collector version is the 74LS09. The 'open collector' and 'open drain' descriptions just mean that the upper transistor in the totem pole is missing. Simply taking away the upper transistor would reduce the output to the equivalent of a single switch, as in Figure 11.7.

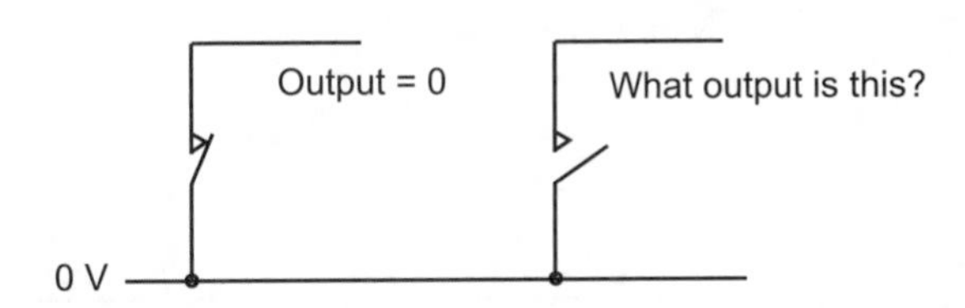


**Figure 11.7**
Open collector gates

The left-hand situation is clear enough; the switch is closed and the output is connected to the 0 V supply. The right-hand one is intended to be at logic 1. However, the output is actually disconnected and is said to be 'floating'. To make sure that the output does go to logic 1 when disconnected, we have to insert a resistor between the output and the +5 V supply.

This resistor is shown in Figure 11.8, and is called a 'pull-up' resistor.

The pull-up resistor is a separate resistor that we have added to the outside of the gate. Its value is somewhere around 1 kΩ, so start with this value and see what happens.

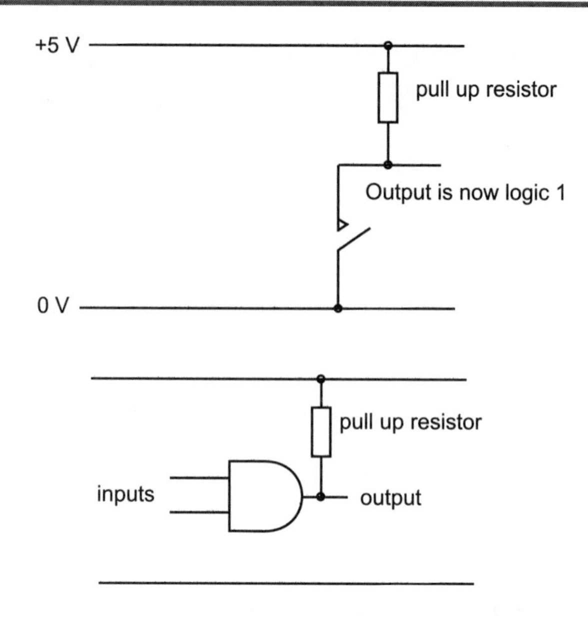


**Figure 11.8**
Pull-up resistors

A single pull-up resistor can serve several open collector gates (Figure 11.9). This can be done to drive an external circuit where the supplies are higher than the normal logic supply. Just remember to check in the data sheet to see the maximum voltage that can be connected to the gate output.

The disadvantage of open collector devices with pull-up resistors is that they cause the circuit to run slightly more slowly and introduce some additional noise.

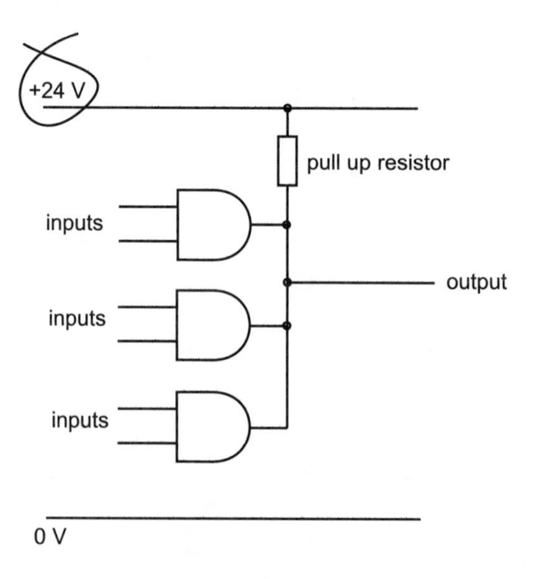


**Figure 11.9**
A useful way of controlling an external circuit

## Power supplies

Mains powered DC supplies are designed as a fixed voltage supply or a variable voltage supply. The fixed ones are OK so long as you choose the right one, but some variable supplies can cause problems.

There are two dangers with these regulators. First, it is very easy to switch on and THEN set the voltage and, if we are unlucky, the voltage could be set too high and the circuit could be destroyed before we have time to adjust it. It is not even safe to reduce the supply to zero and then connect it before increasing it to the operating value. The danger here is that the supply is always connected before the inputs to the gates are applied. Second, many variable supplies create a spike at the moment of switching on, which can exceed the maximum rating of the digital circuit (Figure 11.10).

**Figure 11.10**
Beware of spikes

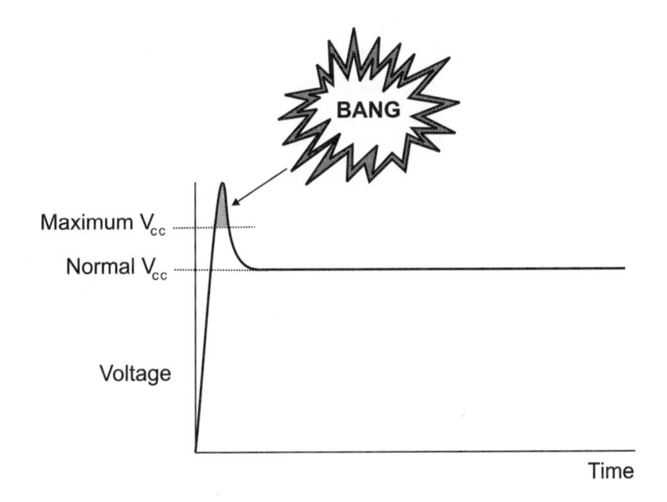


If an unknown variable supply is to be used, the safest way is to:

1 Disconnect inputs from all gates.
2 Disconnect the variable supply.
3 Switch on the supply and adjust it to the correct value.
4 Connect the digital circuit.
5 Connect the external circuits.

### Floating inputs

To prevent accidental switching of circuits, unconnected inputs called 'floating inputs' should be tied to logic 1 or logic 0 as appropriate for the circuit action. A direct wire or track can make the connection on the printed circuit board, provided that the link is less than about 30 cm in length. If it is longer than this it can act as an antenna and pick up noise from the surrounding circuits. CMOS ICs are particularly prone to problems with floating inputs, and the circuit will prove very unreliable if they are left unconnected. TTL inputs tend to float high – that is, they appear to be high if the logic state is measured – and it is very tempting to say that since it is high anyway, it does not need to be tied high. Don't believe it. It will be a constant source of random failures.

## Decoupling

When a gate switches, it goes through a brief moment when both the upper and lower switches in the totem-pole output are partially ON. This causes a short duration surge of current at the moment of switching. The size of this extra current flow depends on the family in use. The 74AC family can create a 50 mA spike. This extra current flows through the conductors on the printed circuit board and causes a momentary drop in the value of all the chip supplies, as in Figure 11.11. A design point here. When we are designing the printed circuit board, we can combat the effect of this current by reducing the resistance of the supply track to the chips. If we make the power supply tracks wider, the resistance falls and, by Ohm's law, the voltage drop is less. A smaller spike is easier to deal with.

**Figure 11.11**
Spikes

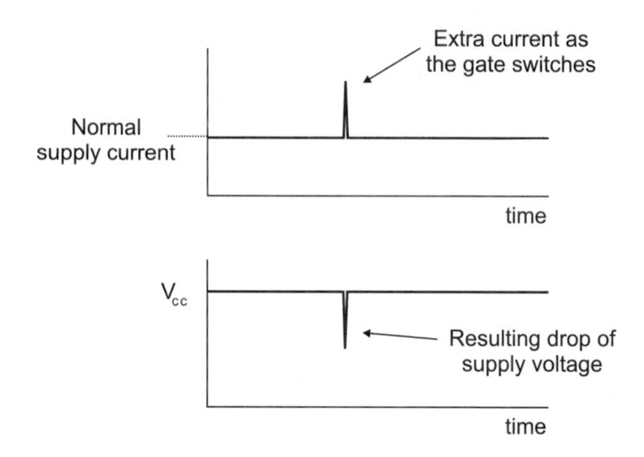


In digital circuits many ICs may switch at the same moment, causing severe spikes, and the resulting drop in supply voltage can cause some logic 1 levels to fall to logic 0 just for a moment. This can cause all sorts of mischief in a logic circuit and can be extremely difficult to track down.

## What size of decoupler?

The spike can be greatly reduced by connecting a small capacitor called a decoupler something like 1–10 nF or so between the chip supply $V_{CC}$ and the 0 V input.

The capacitor acts as an emergency supply of charge to supplement the real supply when the required current increases rapidly. How much charge needs to be held by the capacitor depends on the amount of current drawn by the spike and its duration.

## How many decouplers?

The number of decoupling capacitors fitted depends on the severity of the likely spikes. That, in turn, depends on the logic family used and the number of gates that are likely to switch at the same moment. The other consideration is the possible result of not having enough decouplers. This is a matter of opinion; we would not be too troubled by a Mickey Mouse keyring that fails to function occasionally, but if we were taking off in an aircraft we would like to think that the control circuitry had bucket-loads of carefully selected decouplers.

The number of decouplers varies. The tendency nowadays is to add a decoupler across each IC if it is at all critical, and between every two or five chips in non-critical situations.

The decouplers should be connected between $V_{CC}$ and ground and as close as possible to the IC.

## Ringing

The connections on a printed circuit board have distributed capacitance and inductance. This means that they have a resonance frequency and, given the chance, will oscillate. The resistance will cause any oscillations to be damped – that is, to decrease in amplitude until they die away – but any sudden switches in level will cause such a damped oscillation. After all, switching is just what a digital circuit is designed to do. The ringing causes overshoots when the voltage switches, as seen in Figure 11.12.

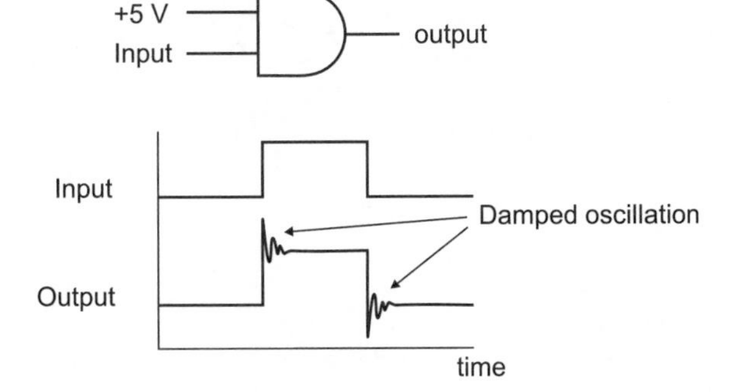


**Figure 11.12**
Ringing

## Crosstalk

When any conductor passes a changing value of current, the surrounding magnetic field changes in sympathy. Faraday's law tells us that any conductor being cut by this changing field will have a voltage induced in it. We are used to this idea with transformers and generators, but often overlook the same situation on a printed circuit board. The sudden switching of a gate with its associated ringing will cause a signal to be induced into nearby conductors. This effect is called 'crosstalk', and can cause accidental inputs to surrounding gates. See the effects in Figure 11.13.

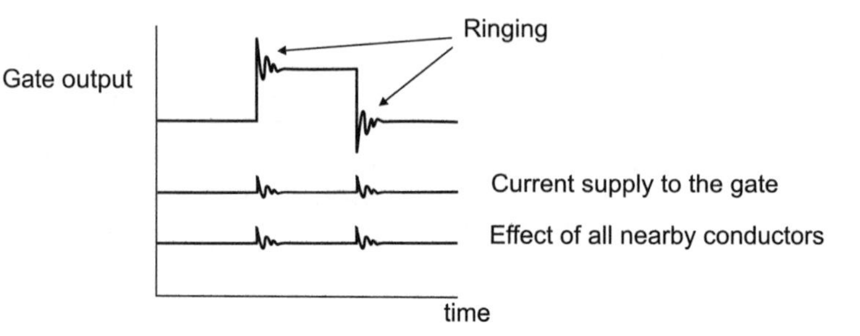


**Figure 11.13**

Crosstalk

## Slow input signals

You may remember the diagram we used to define the slew rate (Figure 10.6). This was stretched out to emphasize that the rising and falling edges were not exactly vertical.

If we allow the voltage to increase slowly, it will reach a stage where internal noise causes the output to switch between levels. The result looks like a burst of high frequency oscillation that occurs at both the input and the output, as seen in Figure 11.14

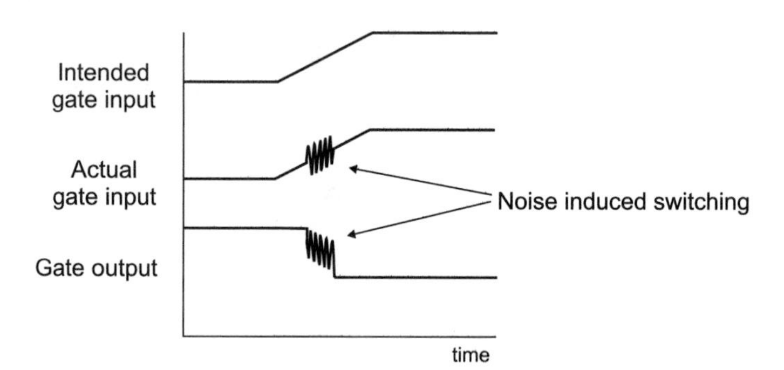


**Figure 11.14**
Low slew rate problem

## Schmitt triggers

These circuits are designed to overcome this problem of slow rising and falling edges. If a Schmitt trigger has a slow rising voltage applied, the output will stay at a low level until it decides to change, and when it does change it goes really quickly. The same performance occurs with a decreasing voltage. The logic 1 at the output is unaffected and then suddenly changes. These can be built from separate transistors, from an operational amplifier or bought ready-made in a 74XX132 chip.

They are also very useful when we have several circuits that are to be switched at the same moment but are activated by slightly different voltages. A low slew rate will cause a significant loss of synchronism between the circuits.

See the benefits in Figure 11.15.

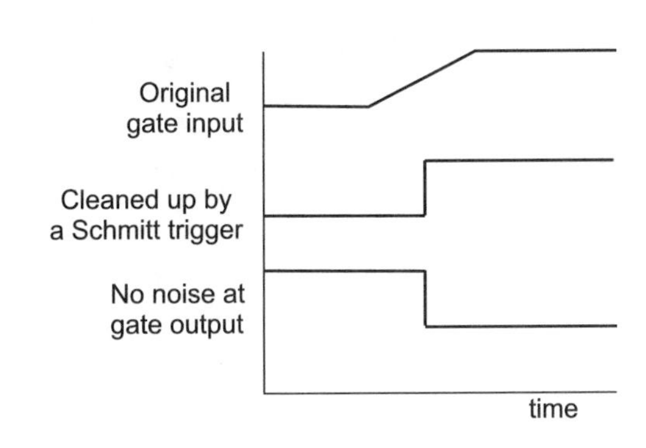


**Figure 11.15**
Schmitt trigger to the rescue

## Quiz time 11

In each case, choose the best option.

**1 The most likely way in which one gate can affect another is by:**

(a) ringing.
(b) opening its collector.
(c) crosstalk.
(d) decoupling.

**2 Spikes can be the result of:**

(a) a sudden opening of the drain.
(b) the discharge of a decoupling capacitor.
(c) the switching sequence of a totem-pole output.
(d) too high a value of fanout.

**3 A floating input:**

(a) can cause a circuit to behave erratically.
(b) causes decoupling of the output.
(c) occurs when the input to a gate is connected accidentally to the ground.
(d) could not occur in a single input gate like an inverter.

**4 A pull-up resistor is essential with:**

(a) an open drain gate.
(b) a high-speed circuit.
(c) a TTL gate.
(d) a variable power supply.

**5 Switch bounce can:**

(a) eliminate the need for decoupling capacitors.
(b) cause a gate to switch many times.
(c) allow a circuit to use a higher supply voltage.
(d) be eliminating by avoiding rubber in their construction.

# 12

# Sequential logic

## It's time to move on

So far all our logic circuits have been simple in the sense that we have some gates, apply some inputs, and we can say with certainty what the result will be. This type of logic is called 'combinational' logic.

The difference with sequential logic is slight but significant. To be able to predict the output from a sequential logic circuit, we need to know the input conditions and, in addition, the previous state of the circuit.

This situation is very common in real life. My television has an on/off button. The first time it is pressed the television is switched ON, but the next time it is pressed the television goes OFF. To make a simple idea more impressive, we could call this a sequential operating system. If someone were to ask us what would happen if we pressed the button, we could not answer unless we knew the current state of the television.

## Latching and non-latching switches

Push buttons and other switches come in two flavours; latching and non-latching.

A doorbell is non-latching – the bell rings only while the button is pressed. A latching switch, as in our television, has a form of ratchet action. It switches states and then stays like that until it is operated again, otherwise we would have to keep our finger on the button all through the programme.

## Timing diagrams

In sequential logic the timing of events is going to be important. To analyse the action of a sequential logic circuit, we have to know both the truth table and also the order in which events occurred. In Figure 12.1 both buttons are pushed after 4 seconds and released 2 seconds later. The latching button changes immediately and holds its value, whereas the non-latching, or 'momentary' switch drops back to 0 V as soon as it is released.

**Figure 12.1**
A timing diagram

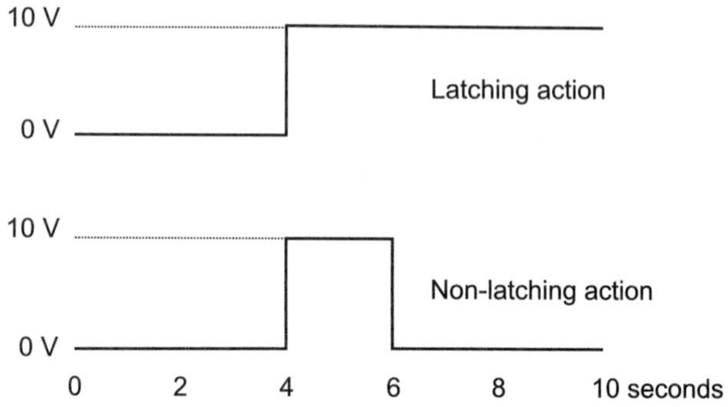

10 V
0 V
Latching action
10 V
0 V
Non-latching action
0 2 4 6 8 10 seconds

Figure 12.2 shows a real logic circuit timing diagram. In this case we are using a NOR gate to invert a continuous square-wave signal.

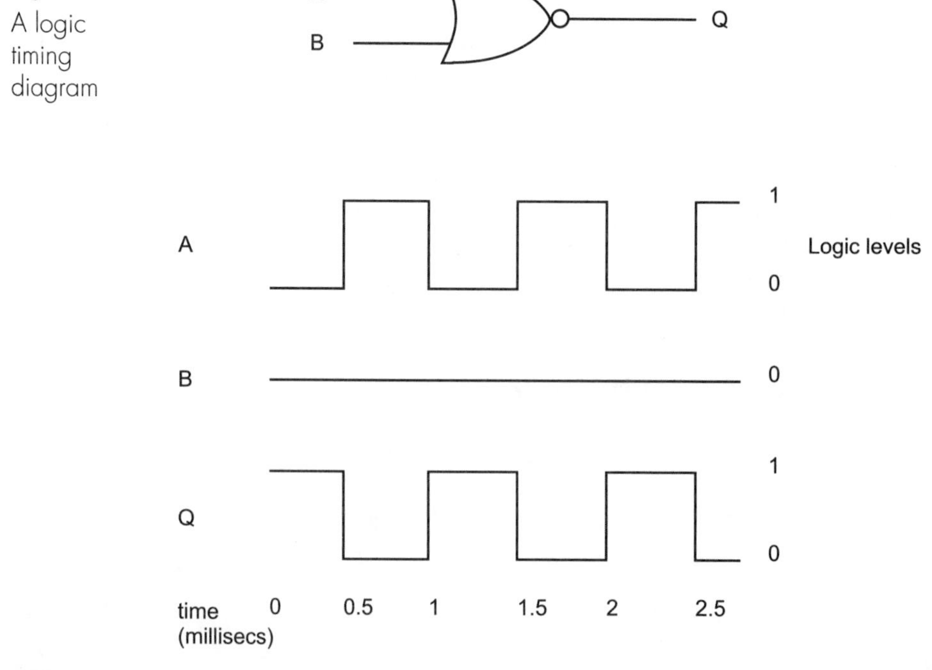


**Figure 12.2**
A logic timing diagram

The basic building block of sequential logic circuits is the bistable circuit. As the name suggests, this circuit has two stable states, which means that the output can remain at a logic 1 level or it can remain at a logic 0 level until an input signal tells it to switch states.

## The RS bistable

There are only three basic bistables in use, and we will start with one that we can build ourselves from a couple of NAND gates. The others are discussed in Chapter 13.

This little circuit goes under a wide variety of names. A bistable is also called a flip-flop on account of its two stable states, or the way it can flip or flop from one state to the other. It can also be called a 'latch', just like the latch switch.

This particular version is an RS bistable or reset/set bistable. The word 'set' refers to the process of making the output go to a logic 1, and 'reset' means the output goes to logic 0.

With so many variations in terms we have to be a little flexible about the names that we meet. The RS bistable is also called: the SR bistable, RS flip-flop, SR flip-flop, RS latch and the RS flip-flop.

The logic symbol for an RS bistable is shown in Figure 12.3. In nearly all sequential logic ICs there are two outputs, one of which is just an inverted version indicated by the circle shown at the output. In some diagrams this inverting circle is not shown, and the fact that the two outputs are labelled Q and $\overline{Q}$ is taken to indicate the inversion.

**Figure 12.3**

The symbols for an RS bistable

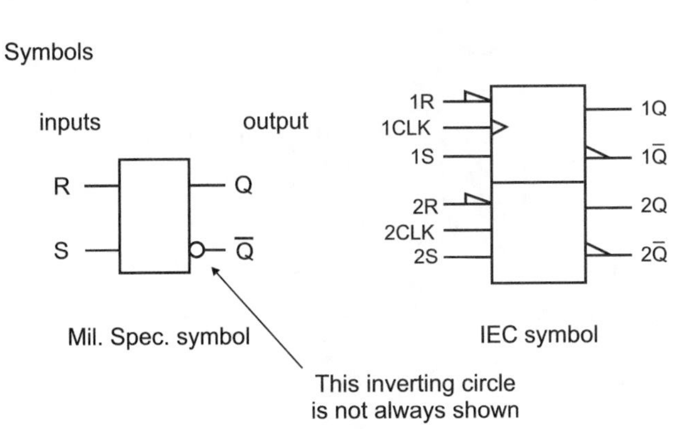


We can buy an integrated version listed as a 'quad SR latch' and numbered 74XXX279. The 'quad' just means that we get four in each IC. As an alternative, we can build them ourselves either from NAND gates or from NOR gates. Since the NAND gate is the more popular version, we will choose this one to investigate.

## The NAND latch

The NAND latch is just a pair of cross-coupled NAND gates, as in Figure 12.4.

**Figure 12.4**
The NAND latch or RS bistable or . . .

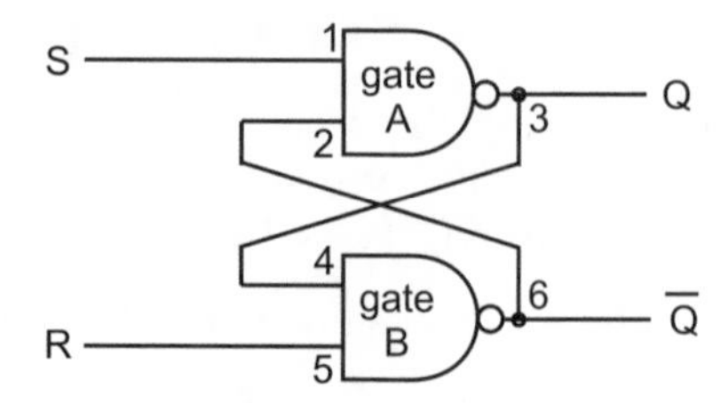


This looks a bit of a puzzle at first, so we will jump straight in with the inputs set as follows.

### S = 1, R = 0

The inputs and outputs are numbered as shown in Figure 12.4.

When input S (1) is at logic 1 the output (3) could be either logic 0 or 1, depending on the state of the other input at (2). This is not very helpful, so we will take a look at R, the input to the other gate (5).

With the reset input (5) at logic 0, its output at (6) is definitely at logic 1 because of the truth table for the NAND gate. Since (6) is connected to (2), gate A will now have both its inputs at logic 1 and this will result in its output (3) being at logic 0. This would make gate B input (4) to be at logic 0.

The final state is the Q output from gate A at logic 0 and the NOT Q, or $\overline{Q}$, output at logic 1.

The first line of the truth table will be as in Figure 12.5, showing that the bistable is 'reset' because Q = 0.

**Figure 12.5**
The NAND latch is reset

| inputs | | outputs | | |
|---|---|---|---|---|
| S | R | Q | $\overline{Q}$ | Comments |
| 1 | 0 | 0 | 1 | reset |

### S = 0, R = 1

The reasoning behind the result of these inputs follows the same pattern as in our last example. Briefly, if S = 0, the output at (3) must be logic 1. Gate B has two inputs, both at logic 1. Point (5) is the R value and (4) is connected to (3), and so its output (6) must be at logic 0.

The final result is S = 1 and R = 0 gives Q = 1 (and $\overline{Q}$ = 0, of course). The bistable is said to be 'set' (Figure 12.6).

**Figure 12.6**

Q = 1 means it is 'set'

| inputs | | outputs | | |
|---|---|---|---|---|
| S | R | Q | $\overline{Q}$ | Comments |
| 1 | 0 | 0 | 1 | Reset |
| 0 | 1 | 1 | 0 | Set |

### Active low inputs

If we wish to 'set' the latch, i.e. make the Q value go to logic 1, we must apply a logic 0 to the set input. Similarly, if we wish to reset it, we must apply logic 0 to the reset connection. To perform their stated function, either S or R, we must apply logic 0. These inputs are therefore referred to as 'active low'. On logic diagrams, the active low characteristic is indicated by putting a 'not' line over the input – in our case, S and R.

By a small majority, most inputs to most ICs are active low.

### S = 0, R = 0

What would happen if we tried to set and reset the latch at the same time? Let's see.

By taking S to a logic 0, gate A will make its positive output go to logic 1 owing to the nature of the NAND gate.

Similarly, by making R = 0, gate B will have an output of 1. It is a stable condition and does no harm to the gates, but we avoid this situation and refer to this state as 'forbidden', 'prohibited', 'invalid' or 'indeterminate'. This is shown in Figure 12.7.

**Figure 12.7**

The 'forbidden' state

| inputs | | outputs | | |
|---|---|---|---|---|
| S | R | Q | $\overline{Q}$ | Comments |
| 1 | 0 | 0 | 1 | Reset |
| 0 | 1 | 1 | 0 | Set |
| 0 | 0 | 1 | 1 | Forbidden |

Why don't we like this state? Simply because we have labelled the two output as Q and $\overline{Q}$, which by our definition of Boolean means that the two outputs have opposite logic states; however, in this 'forbidden' state we have Q and $\overline{Q}$ with the same logic state. It therefore contradicts what we said it would do.

### S = 1 and R = 1

Assume we have the 'set' condition shown in Figure 12.8.

Now, what would happen if S changed to logic 1? The logic level at output (6) is still 0, and therefore the input to gate A at (2) is still 0. As we said earlier, if one input to a NAND gate is logic 0 the output will be at logic 1 regardless of the other input to the gate.

**Figure 12.8**
What happens if S goes to logic1?

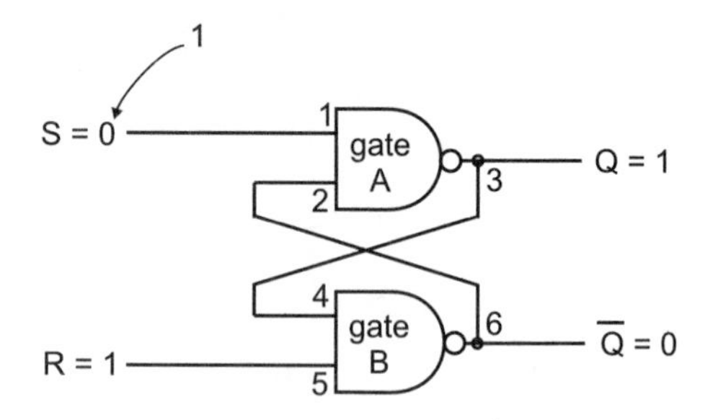


The answer to our question is . . . nothing. If S went to logic 1 there would be no change in the output.

If we had a latch with S = 1, R = 0 giving Q = 0, changing R to 1 would not change the output state at all.

Changing either input to logic 1 has no effect on the output state, and the circuit is said to be in a 'memory' or 'latch' state.

The final truth table is shown in Figure 12.9, now following the convention of listing the inputs in binary order.

**Figure 12.9**
The complete NAND latch truth table

| inputs | | outputs | | |
|---|---|---|---|---|
| S | R | Q | $\overline{Q}$ | Comments |
| 0 | 0 | 1 | 1 | forbidden |
| 0 | 1 | 1 | 0 | set |
| 1 | 0 | 0 | 1 | reset |
| 1 | 1 | Q | $\overline{Q}$ | memory or latch |

## A use for a NAND latch

Let's see how we could use a bistable. Imagine that we have a circuit running from a 5 V supply that fails every hour or so, and we feel that there is a bad joint somewhere that is causing the input voltage to be disconnected for a very brief moment. Extremely short and difficult to detect pulses are called 'glitches', and the circuit designed to detect them is called a 'glitch catcher'.

One way of finding out whether or not this power supply failure occurs is by using a NAND latch to monitor the voltage.

In the circuit shown in Figure 12.10 we would prepare the trap by connecting the S input to the suspect power supply and R to a known good supply. Now, we reset the latch by closing switch SW1 for a moment to make the output change to Q = 0 and Q = 1. By reopening the switch we have S = 1 and R = 1, which is the memory state, and the outputs would be held at their present values.

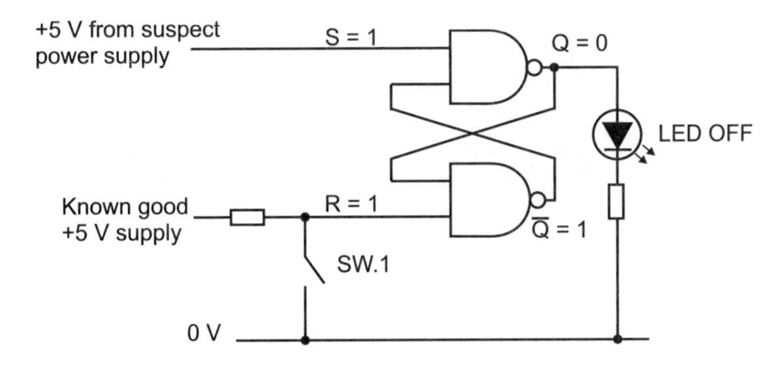


**Figure 12.10**
The latch is on guard duty

The monitor LED would be unlit because the Q output would be at logic 0, so there would be no voltage across it.

The trap is prepared, so all we have to do is to wander off for an extended lunch and see what happens.

If the fault occurs, the S input will drop to 0 V just for a moment. As it is an active low circuit, the latch will be set and Q will go to logic 1. As soon as the power supply recovers the latch will have inputs of S = 1 and R = 1, which is the memory state, so the output will not change.

Since the Q output has gone to logic 1 it will provide the current to switch on the LED as in Figure 12.11.

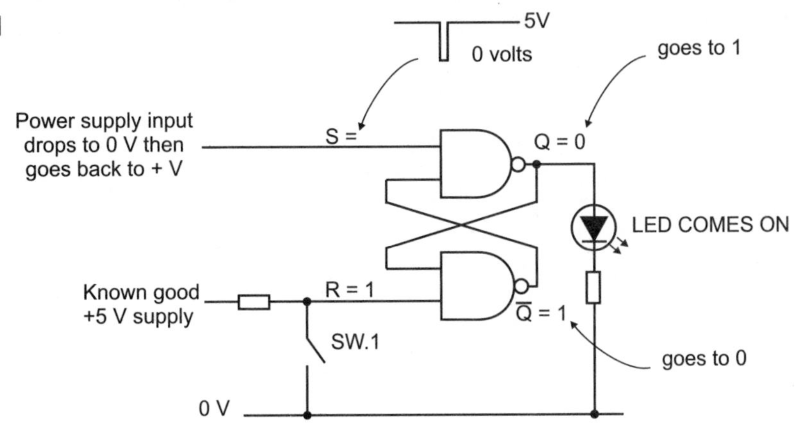


**Figure 12.11**
The fault is detected

## Another use – a cure for switch bounce

We have mentioned this one in the previous chapter. This circuit uses a changeover switch, often called a single-pole double-throw switch.

As the switch changes position, it will be connecting logic 0 input to one or other of the gates. This will cause the output voltage to change state. If the switch bounces the contact will disconnect and both inputs will be returned to logic 1 via the pull-up resistors, and so the latch will go to its memory state and hold the previous outputs. Hence no bouncing. This circuit is shown in Figure 12.12.

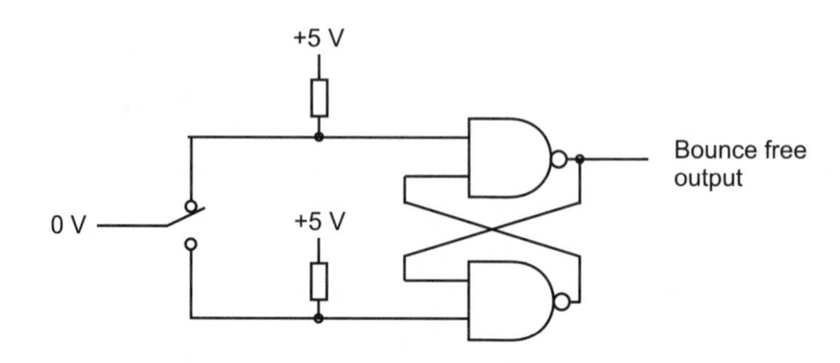


**Figure 12.12**
Switch debouncer

## A NOR latch

Although less popular, the same functions can be performed with a couple of NOR gates. As a result of the differing truth tables, the detailed operation in terms of logic levels works out slightly differently. The circuit diagram and the truth table are shown in Figure 12.13. The inputs to the NOR latch are active low, just as in the NAND gate, so it sets (Q = 1) when S goes to logic 0.

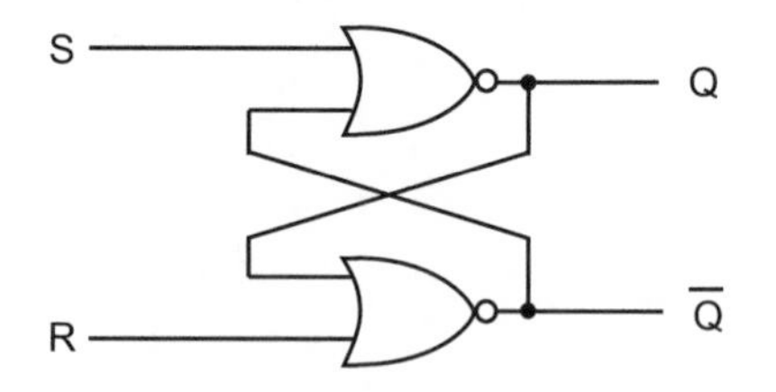


**Figure 12.13**
The NOR latch and truth table

| inputs | | outputs | | |
|---|---|---|---|---|
| S | R | Q | $\overline{Q}$ | Comments |
| 0 | 0 | Q | $\overline{Q}$ | memory or latch |
| 0 | 1 | 1 | 0 | set |
| 1 | 0 | 0 | 1 | reset |
| 1 | 1 | 0 | 0 | forbidden |

## Testing the glitch catcher

If we built a latch and wanted to see if it could detect very short duration changes of logic state, our first problem would be getting hold of a pulse less than, say, 50 ns wide (1 ns = $1 \times 10^{-9}$ s).

Easy! Use a switch.

Snag! It will probably bounce.

Easy! Build a debounce circuit or buy a bounce-free switch.

Snag! A switch won't be fast enough – it can manage 10 ms at best.

We cannot use a switch, so we need something faster – like a logic gate.

### Transition effects

If we apply a changing input to a NOT gate the gate will just invert the input signal, so if we plotted the output at 1-s intervals the timing diagram would look as in Figure 12.14.

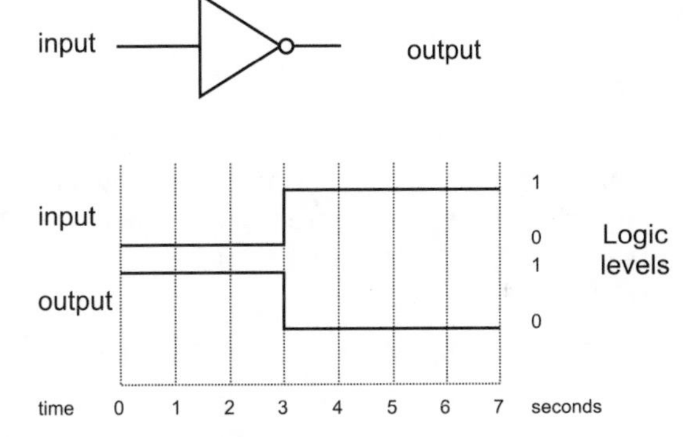


**Figure 12.14**
Switching a NOT gate

This figure is sort of, mostly, true. If we take this same situation but look at it more closely by plotting it at 10-ns intervals around the time of switching, we get the effect shown in Figure 12.15.

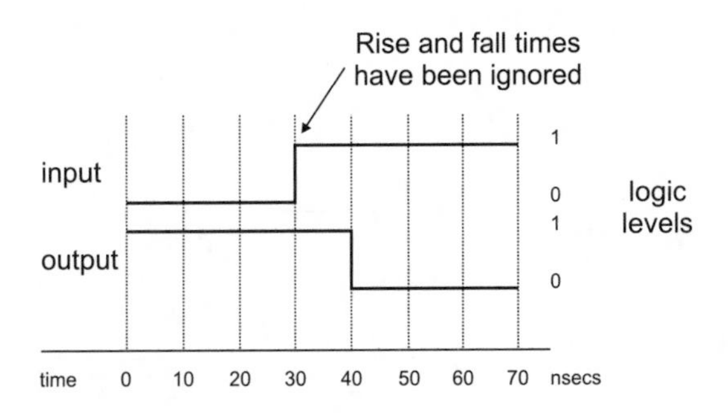


**Figure 12.15**
A close look at the action

A curious situation has occurred. The change at the output is delayed by the time it takes the voltage to travel through the gate – called the transition or propagation time. For a brief moment, about 10 ns, the input and the output of the gate have the same logic level.

So, does it matter?

Not always, but sometimes it can cause unexpected effects.

Have a look at Figure 12.16 and see what output, if any, you would expect to get from this circuit.

**Figure 12.16**

Is there an output when the switch closes?

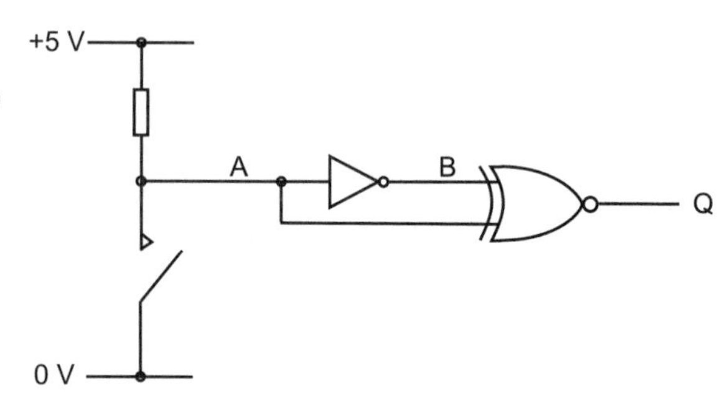


The simple view is that the inputs to the XNOR gate are always opposite in value, since one of the inputs is always passed through an inverter. As we know from earlier work, an XNOR gate has logic 1 output whenever the two inputs are the same. So there should be no output.

In reality we have seen that the inverter will delay the signal by a few nanoseconds, and during this time the two inputs to the XNOR gate will have the same values. This will result in logic 1 appearing at the output of the circuit for about 10 ns. See the waveforms in Figure 12.17.

**Figure 12.17**

A home-made glitch

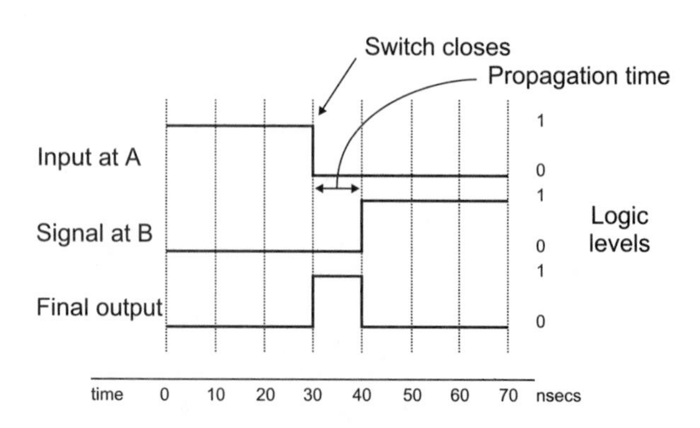

Much the same effect can be obtained with other gates – there was no special reason why we used an XNOR gate in our example.

This output pulse is too fast to be seen on most of our test instruments, so a latch or something else must be used to detect it.

If we want a longer pulse, we can just add a larger number of NOT gates.

## Static hazard

We have seen that small time delays as the signal passes through a part of the circuit can cause unexpected voltage spikes. This effect is called a static hazard. To avoid them, we must balance the time delays on the gate inputs to ensure that voltage changes occur at the same time. This may even require us to pass the signal through another gate just to introduce a time delay.

The moral of this chapter is that if we have a logic circuit that is behaving oddly, things may be happening that are not obvious at first or even second glance.

## Quiz time 12

In each case, choose the best option.

**1 On the number pad of a telephone the switches are called:**

(a) 'momentary' switches.
(b) latching switches.
(c) active switches.
(d) ratchet switches.

**2 A NAND latch has a memory state whenever:**

(a) inputs S and R are set to logic 0.
(b) both outputs have the same value.
(c) inputs S and R are set to logic 1.
(d) the inputs are active low.

**3 To 'set' a NAND latch, the logic state of the R input:**

(a) must be logic 1.
(b) must be the same as the S input.
(c) must be logic 0.
(d) does not matter, it is only the S input that matters.

**4 A glitch is a:**

(a) person who is obsessed with computers.
(b) cause of switch bounce.
(c) forbidden state in a latch truth table.
(d) momentary change of logic level in a signal.

**5 In a NAND RS latch the S input is held at logic 0 while the R input changes from logic 1 to 0, the latch will:**

(a) go to its forbidden state.
(b) be cleared.
(c) enter its memory state.
(d) be damaged.

# 13

# Clocked bistables

## Clocks

All gates and the SR bistable operate immediately when we apply the inputs. The technical term for this sort of 'go when you like' circuit is an asynchronous operation.

We don't always want this sort of operation. Much of our lives is governed by the idea of gathering all the requirements for a job but not actually starting it until a signal is given. We don't want our alarm clock to ring immediately we have set it. We don't want the plane to take off as soon as the pilot arrives at the airport, nor do we want the chef to start cooking as soon as the first ingredient is delivered, whatever the time of day or night. This 'wait until you are told' routine is called synchronous operation.

Most digital circuits are synchronous. An obvious example is a digital clock, in which a circuit counts the seconds up to 60 and then sends a pulse to the minute counter to increase its value by one. The 'seconds' counter is also reset to zero to restart the counting process.

Let's see how to convert the RS bistable to a clocked RS bistable.

## Clocked RS bistable

The only change we have to make is to add a 'wait until you are told' circuit. We can do this by adding a couple of NAND gates. Have a look at Figure 13.1.

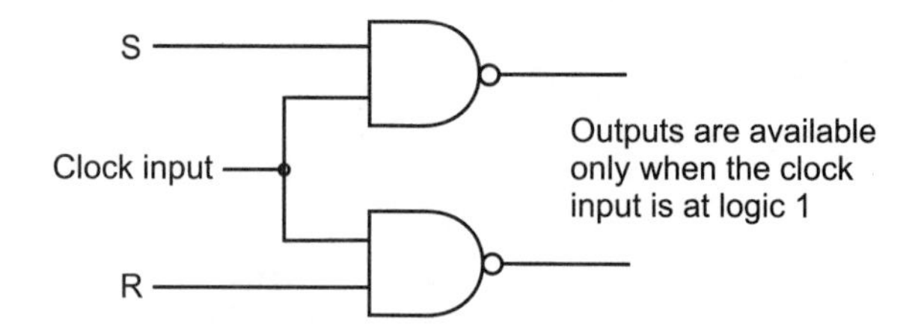


**Figure 13.1**
The inputs can be clocked

If the clock input signal is held at logic 0, the NAND gates are switched off and both outputs are at logic 1. Notice how the values of S and R will now have no effect on the output.

When the clock input goes to logic 1, the S and R inputs appear at the output and are applied to the bistable.

The complete circuit and truth table are shown in Figure 13.2. Notice how the logic levels of the S and R inputs are inverted by the NAND gates in the clock circuit, and this has a knock-on effect on the truth table.

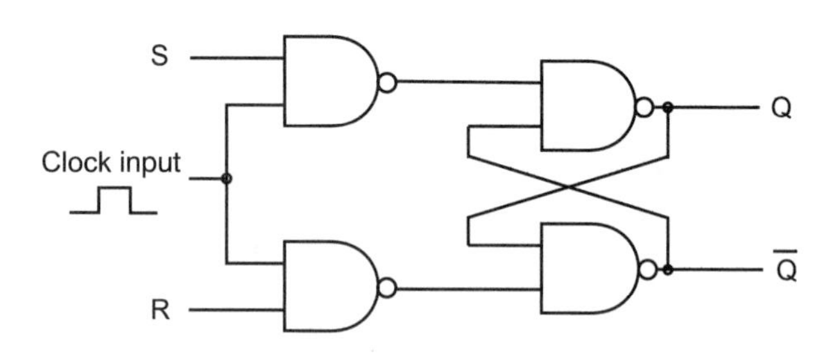


**Figure 13.2**
The clocked SR bistable

| inputs | | outputs | | |
|---|---|---|---|---|
| S | R | Q | $\overline{Q}$ | Comments |
| 0 | 0 | Q | $\overline{Q}$ | memory or latch |
| 0 | 1 | 0 | 1 | reset |
| 1 | 0 | 1 | 0 | set |
| 1 | 1 | 1 | 1 | forbidden |

The advantage of using NAND gates throughout this circuit is that we buy NAND gates four at a time in the 74XX00 chip. So for a few pence we can build the whole circuit using only one component.

## The JK bistable

This is the second of the four bistables that are available. It is particularly popular because we can easily make it behave like any of the others. For this reason it has been referred to as the 'universal' bistable.

While it is quite possible to build a JK from basic gates, or even from transistors, it is a pointless exercise. In the integrated form, we can buy two JKs in a single chip for less than a cup of tea. To build it from separate gates or components would not only be more expensive but it would be more difficult, consume more power and take more space.

In many ways the JK bistable is very similar to the clocked RS bistable. In fact, all the bistables are only slight modifications of each other.

There are two data inputs called J and K, which are similar to the S and R inputs that we have met. The JK has two outputs, Q and $\overline{Q}$, just as before. It also has a clock input.

One improvement over the SR bistable is that the 'forbidden' state has been replaced by something useful. This is called a 'toggle' state. This means that the Q and $\overline{Q}$ outputs alternate between logic 0 and 1 each time a clock pulse arrives.

The logic symbol and truth table are shown in Figure 13.3. Unfortunately the symbols on truth tables have not been standardized, so we have to be a little flexible when we interpret them. Different manufacturers present them differently.

**Figure 13.3**
The JK bistable

Symbols

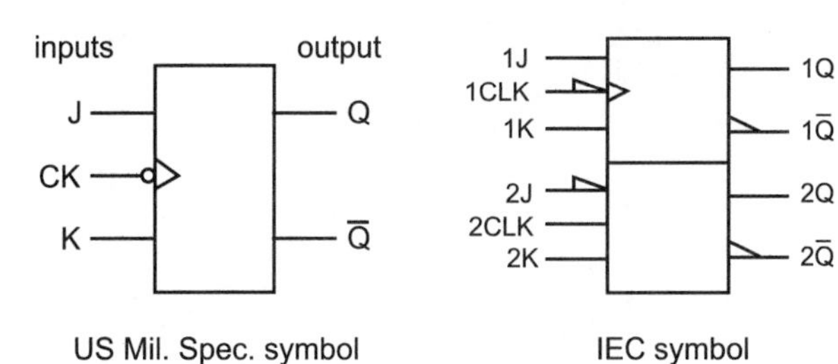


US Mil. Spec. symbol

IEC symbol

| inputs | | output | |
|---|---|---|---|
| J | K | $Q_{n+1}$ | Comments |
| 0 | 0 | $Q_n$ | memory or latch |
| 0 | 1 | 0 | reset |
| 1 | 0 | 1 | set |
| 1 | 1 | $\overline{Q}_n$ | toggles |

On this truth table:

$Q_{n+1}$ = the next output state, that is, what will happen after the clock pulse has occurred.

$Q_n$ = the present output state before the clock was applied.

$\overline{Q}_n$ = the opposite output state to the one that was present before the clock was applied.

We must remember that if no clock is applied, then nothing happens.

**The golden rules of JKs**

**1 If J and K are different, the Q is always the same as the J.**
**2 If J and K are 0, nothing happens.**
**3 If J and K are 1, the output toggles.**

## When does the clock tick?

The clock input determines the moment at which the circuit will respond to the state of the inputs. There are three possibilities. The one in use is always stated on the data sheet and shown on the symbol, and has really significant effects on the operation of the circuit. The input signals must be present just before the clock pulse occurs and they sometimes have to be held for a short time afterwards. These times are listed as set-up time, typically 20 ns, and hold time, which is often zero.

## Negative edge triggered

This is also called trailing edge triggered, and indicates that the action occurs at the end of the clock pulse just as its logic level changes from logic 1 to logic 0. This is indicated by the clock input. The > symbol means it is edge triggered, and the circle means the edge in question is the negative-going edge.

Any changes to the input that occur during the rest of the pulse are ignored.

If we apply logic 1 to both the J and K inputs, the JK will toggle on each clock pulse synchronized to the negative edge. So if we apply a continuous square-wave to the clock it has the effect of halving the output frequency, as we can see in Figure 13.4. To keep the figure uncluttered, the NOT Q output is not shown. It is always opposite to the Q output.

**Figure 13.4**
Trailing edge triggering

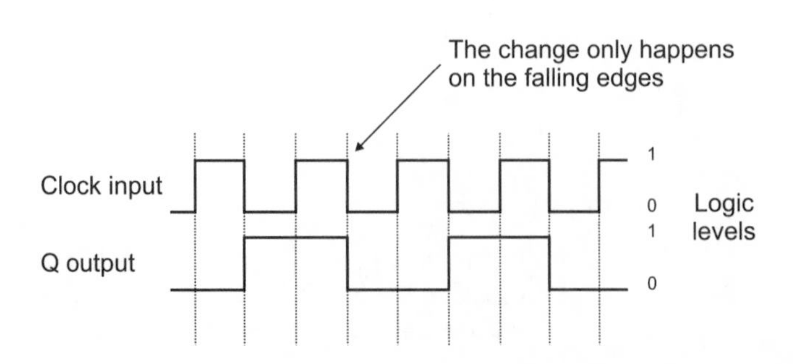

**Positive edge triggering**

This is also called leading edge or rising edge triggering.

Everything is much the same, except that the changes occur on the positive-going edge of the waveform. The symbol will include the > to show it is edge triggered, but this time there will be no circle.

Have a close look at Figure 13.5, comparing the output waveform with that on Figure 13.4.

**Figure 13.5**
Positive edge triggering

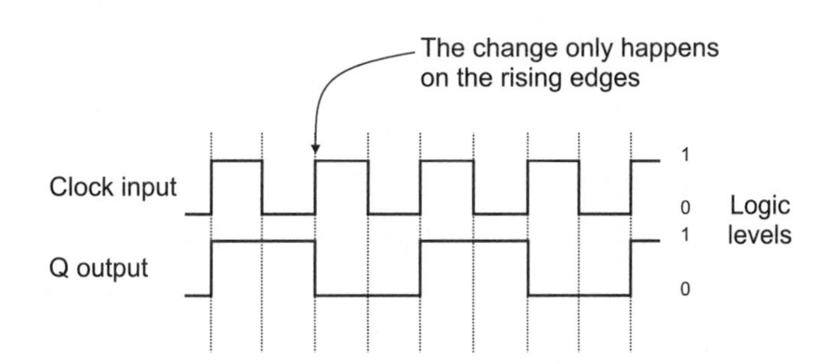


Very few devices are available with positive edge triggering, but that doesn't matter. It is very easy to change a negative edged circuit to use the positive edge if we ever need to. Can you see how we could do it? Have a look at Figure 13.6, and see that we have simply inverted the clock pulse that is being used to supply the positive edge chip. By inverting this input clock pulse, its own positive-going edge coincides with the negative edge of the main circuit.

**Figure 13.6**
Positive edge device pretending to be negative edge

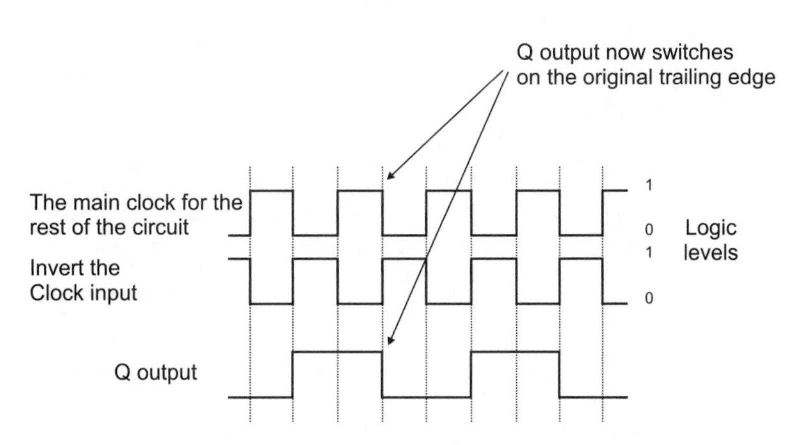


Are there any snags? Yes. One, possibly two.

This clock inversion method will only work if the clock pulse is a square-wave with a mark-space (on–off) ratio of 1, otherwise we have a problem. We are assuming that the waveform is symmetrical and that by inverting it we can have an identical wave but shifted by half a wavelength.

The other question to ask is 'where did we get the inverted wave from?'. If we passed the original clock through an inverter, we will have introduced a time shift of 10 ns or whatever the propagation delay of the chip is. If the clock came from a JK or something similar, we could overcome this problem by using the NOT Q instead of the Q output to provide the inverted version. There is no time delay between the Q and NOT Q outputs of a single bistable.

## Pulse triggering

This is also called a master–slave design.

It changes its state, if required, on the trailing edge of the clock pulse, just like the negative edge triggered bistable, but it actually uses the whole of the input pulse as part of the process.

In the normal trailing edge JK, all the action occurs on the falling edge, inputs are read, output levels are fed out of Q and NOT Q. It's a busy time, and if the input clock is very narrow it's a very busy time indeed. With very narrow pulses being read into a circuit, it is possible to miss a pulse and give the wrong output.

The master–slave design was intended to overcome this possibility by using the main part of the pulse to read the values at the J and K inputs and then, on the falling edge, to pass the appropriate signals out via Q and NOT Q.

The overall action can be likened to a series of barges going through a lock:

1 Both lock doors are closed.
2 The entrance door opens and the lock is loaded with barges.
3 The entrance door closes trapping the barges inside.
4 The exit door opens and the barges come out.

Now, translating this into Digispeak, we have the following action:

1 The JK inputs and the Q and NOT Q outputs are isolated.
2 As soon as the leading edge of the clock pulse reaches logic 1, the data enters and is stored inside the JK.
3 At the end of the clock pulse, when the falling edge leaves the logic 1 level, the J and K inputs are disconnected preventing any latecomers from getting in.
4 The data is transferred to the Q and NOT Q outputs.

See Figure 13.7.

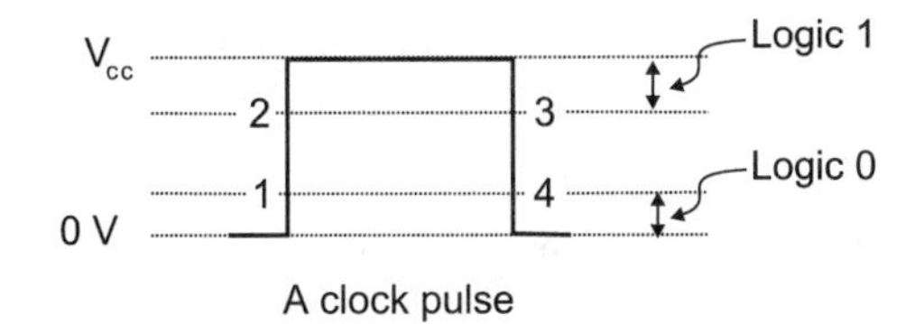


**Figure 13.7**
The steps taken by a master–slave JK

## A master–slave trick

Some master–slave JKs can play a trick that can cause an unexpected output that is difficult to trace. Although both master–slave JKs and negative edge JKs switch on the trailing edge of the clock pulse, they can give different results when fed with the same waveform. That sounds odd.

In Figure 13.8 both JKs have their J and K inputs settled happily at logic 0 before the negative-going edge of the clock pulse. In the case of the edge triggered JK, the logic 0 simply tells the JK to stay in its old state which, in this case has $Q_{out} = 0$.

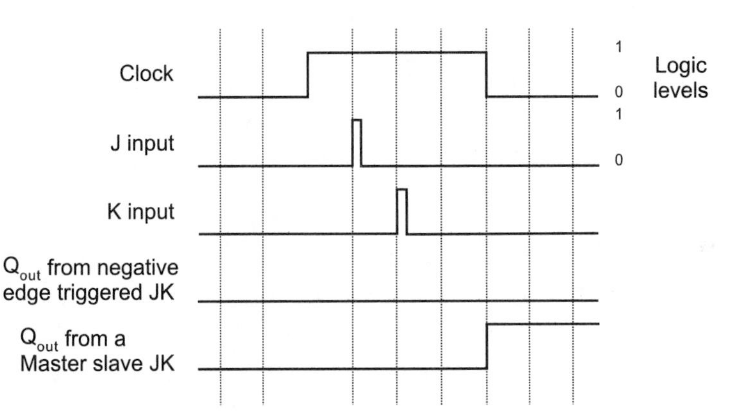


**Figure 13.8**
The great master–slave JK trick

In the case of the master–slave JK, we have allowed the J and K inputs to go to logic 1 for a moment but have safely restored the logic 0 level before the end of the clock pulse. The master–slave circuitry has the ability to remember that one or both of the inputs have been high at some time during the clock pulse.

When the negative edge occurs, the master–slave responds as if the high inputs are still there. In this case, with both J and K apparently high, the JK toggles. This doesn't happen with an edge triggered JK, which simply reads the input values at the time of the clock edge.

## Set and preset inputs

When our computer gets itself into a hopeless muddle and won't respond, the last resort is for us to press the reset button. This button over-rides the program operation in the microprocessor and sets everything back to the starting point.

All bistables and many other integrated circuits have the ability to be reset by over-riding all other inputs and putting the Q output at logic 0. Similarly, many digital ICs can also set the Q output to logic 1.

The reset input is called 'clear', and is written as $\overline{\text{CLR}}$. The line over the top indicates that it is an active low input, so we connect it to logic 0 to reset the bistable. This clear function is called asynchronous, which means that it is independent of the clock input. It sends the output to 0 whenever we like. When the chip is in normal operation, this input must be kept at logic 1.

Some clear inputs are synchronous. That means the outputs do not revert to zero state immediately, but do so on the leading edge of the next clock pulse.

The preset or $\overline{\text{PRE}}$ input works in just the same way, except that in this case it forces the Q output to go to logic 1.

When cleared or preset, the NOT Q output takes up the opposite level to the Q output.

As with all inputs to bistables, the preset and clears must be tied to a logic 1 when not used. Never leave them floating or they may accidentally go low and cause erratic operation of the circuit while they are logic 0, over-riding any other inputs.

## D-type bistable

Now that we have sorted out the JK bistable, the others are very easy. In fact, they are only slight modifications to the basic JK.

Think back to the JK truth table. We have the two cases where J and K are at different logic levels, and two cases where they are the same.

The D-type only uses the first two situations, where J and K have different values. This reduces the truth table to only two lines, and this is the only difference between the D and JK bistables. Simple, isn't it?

How do we manage to ignore two lines in the JK truth table? Simple again; we just make sure that they don't occur. We do this by connecting a NOT gate between the J and K inputs. With a NOT gate between them, J and K can never have the same value.

The finishing touch is to call the input 'D' rather than J, and make the old K input an internal connection so that the D-type bistable has only a single data input. The symbol and truth table are shown in Figure 13.9.

We can buy a D-type or we can connect a NOT gate across the inputs of a normal JK to make our own, just as we wish.

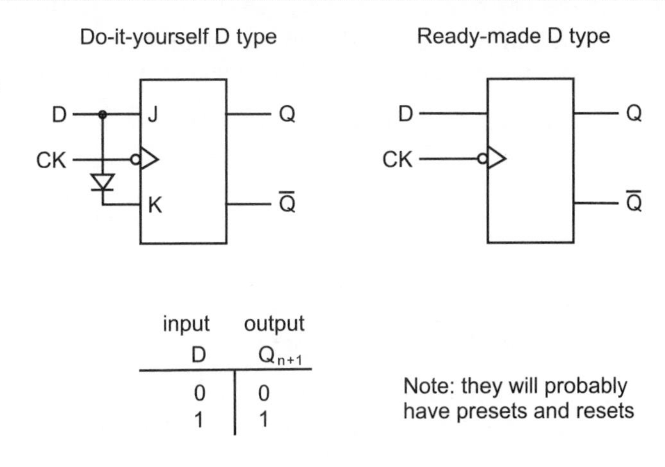


**Figure 13.9**
The D-type bistable

When we see the truth table, it seems a bit pointless. The inputs are exactly the same as the outputs – so why bother? The only benefit is that the output data changes in synchronism with the clock pulse and the rest of the circuit. This bistable is widely used for shifting data around in microprocessor systems, hence the name 'D' for 'data'.

Since the input and output logic levels are always the same, this bistable is sometimes referred to as a 'transparent latch'.

## T-type bistable

T is for toggle. This again is really a JK in disguise. In the JK truth table, we remember that the effect of holding J and K at logic 1 is to make the outputs toggle. The output rectangular waveform runs at half the frequency of the input, just as we saw in Figures 13.4 and 13.5. This is the purpose of this bistable.

Unlike the D-type we cannot buy one readymade, but we don't have to. It is just a JK with the J and K inputs permanently connected or 'hardwired' to logic 1.

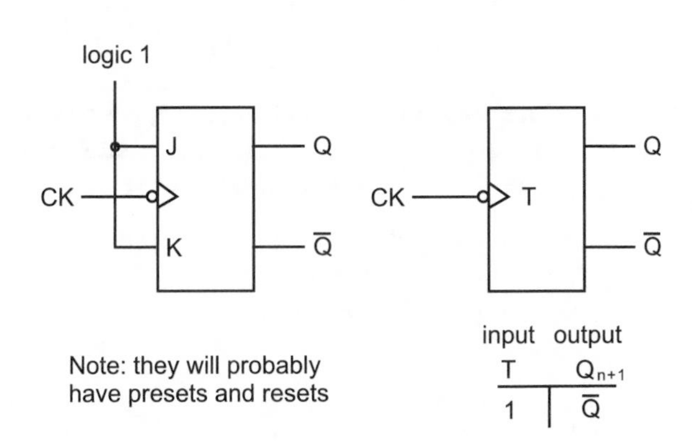


**Figure 13.10**
The T-type bistable

If you were pleased with the simplicity of the D-type truth table, be prepared to be positively delighted when you see Figure 13.10.

That's all of the bistables. In the next chapter we will start to make use of them.

## Quiz time 13

In each case, choose the best option.

**1 A D-type bistable:**

(a) changes its output state after every clock pulse.
(b) has a memory state when both inputs are at logic 0.
(c) doubles the input frequency.
(d) is a clocked bistable.

**2 A JK bistable can be converted to a T-type by:**

(a) adding a NOT gate between the J and K inputs.
(b) connecting both inputs to logic 1.
(c) disconnecting the K input.
(d) applying a square-wave input signal of twice the required frequency.

**3 A JK bistable has inputs of J = 0, K = 1 and Q = 0. After the next clock pulse:**

(a) the NOT Q output would be at logic 1.
(b) the Q output would be at logic 1.
(c) the output is uncertain, it depends on the clock pulse.
(d) the preset input would go high.

**4 The symbol > on a bistable symbol indicates:**

(a) that the bistable is negative edge triggered.
(b) the pin that is to be used as the input.
(c) that the bistable is edge triggered.
(d) the direction of data flow.

**5 The inputs applied to a T-type bistable are T = 0, a square-wave clock pulse, $\overline{CLR}$ = 1 and $\overline{PRE}$ = 0. The Q output will change to:**

(a) logic 1 after the next clock pulse.
(b) logic 0 immediately.
(c) logic 1 immediately.
(d) logic 0 after the next clock pulse.

# 14

# Asynchronous counters

Counters are used to count the number of input pulses occurring in a circuit. The pulses can be derived from radio transmissions, a computer system or, very slowly, the number of coins being fed into a vending machine.

We can build them from JK bistables or D-type bistables. It will not come as a surprise to hear that counters are available in integrated form.

So why build them ourselves if they can be bought readymade?

Three reasons for building our own counter:

1 We need to know the circuitry and possible problems because we will certainly meet them in existing equipment.
2 We can build a counter that meets our design requirements. Like all off-the-peg equipment, standard counters meet standard needs. But we may not be standard.
3 It's fun.

Reasons for using an integrated counter:

1 They are cheaper, easier and quicker to build.
2 They consume less total power.
3 They take up less space.

Integrated counters are discussed in Chapter 15.

## How counters work

As we saw in the last chapter, the output of a JK bistable can be made to toggle if the two inputs are connected to logic 1.

If we look at Figure 14.1, we are reminded that the output is running at half of the frequency of the input clock. We can stay with the idea of a frequency divider, or we can look at it in a different way. We can note that at the beginning of the output cycle arrowed on the diagram the Q output is held at logic 0, and it then changes to logic 1 for the second half of the cycle.

**Figure 14.1**

A divide-by-two circuit

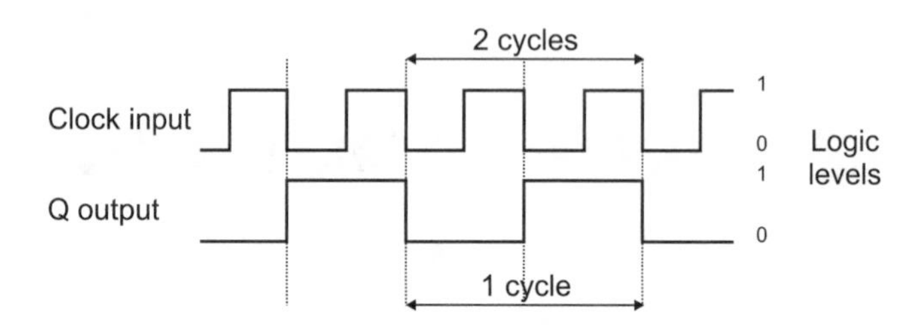


This is a binary count. It counts up 0, 1 and then resets to 0 and starts again. Admittedly it could equally well be described as counting down 1, 0 then resetting to 1, but at the moment this doesn't matter. The thing is, it is either a frequency divider or a counter, whichever we want.

The number of different counts before its starts again is called the 'modulo', so in this case we could describe the circuit as a modulo-2 counter, usually abbreviated to a mod-2 counter.

## Mod-4 counter

If we were to use the output of the mod-2 counter as the clock for another T-type bistable, that bistable would also divide its input by two. Overall, then, the effect would be a circuit that divides the original input by four. We have built a mod-4 counter (Figure 14.2).

We can see the count by reading the values of $Q_0$ and $Q_1$:

1 They start as $Q_0 = 0$, $Q_1 = 0$, and the count is 00.
2 After the trailing edge of $Q_0$, the values become $Q_0 = 1$, $Q_1 = 0$, giving a count of 01.
3 After the next trailing edge of $Q_0$, the values become $Q_0 = 0$, $Q_1 = 1$, giving a count of 10.
4 Then the values become $Q_0 = 1$, $Q_1 = 1$, giving a count of 11. The next count returns to 00, and thereafter it counts continuously.

There is no ambiguity regarding the count here. It is definitely counting up in the binary sequence 00, 01, 10, 11, 00, 01 . . .

**Figure 14.2**
We can now count up to four

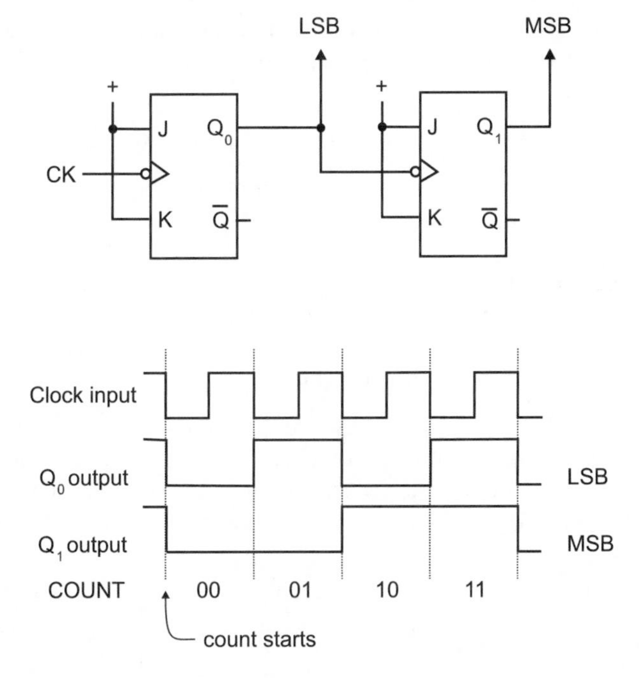


## Mod-8 counter

We could add another bistable on the end of a mod-4 counter and have a mod-8 counter. The count starts from 000 and then proceeds 001, 010 etc. up to 111, then resets to 000. This is shown in Figure 14.3. Another bistable would make it mod-16 and so on; we could go on for as long as we wanted.

**Figure 14.3**
We can add more bistables as required

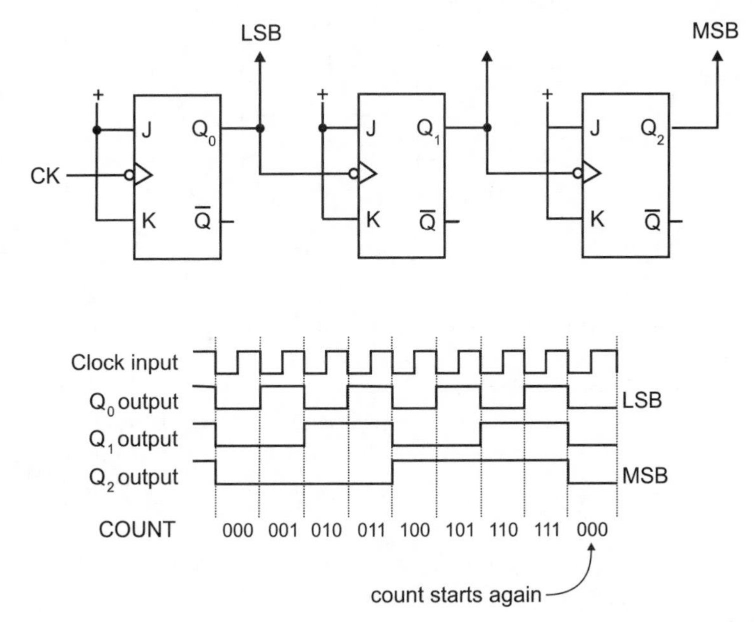

All these counters have been counting up from zero, and we have been using the Q output from each bistable for two functions:

1 Providing the output counts.
2 Providing the clock input to the next bistable.

The NOT Qs have been ignored, so let's see what happens when they join in the fun.

## A few experiments

We will use the mod-4 counter in Figure 14.2 as the test-bed, but we must appreciate that any effects noticed are going to apply equally well to all counters, however many bistables are used in the design.

In Figure 14.4, we leave the circuit alone except for taking the output from the NOT Q outputs rather from the Q output. The effect is that the count now starts at 11 and counts down to 00. It is not difficult, then, to change an up counter to a down counter.

**Figure 14.4**
Using NOT Q to count

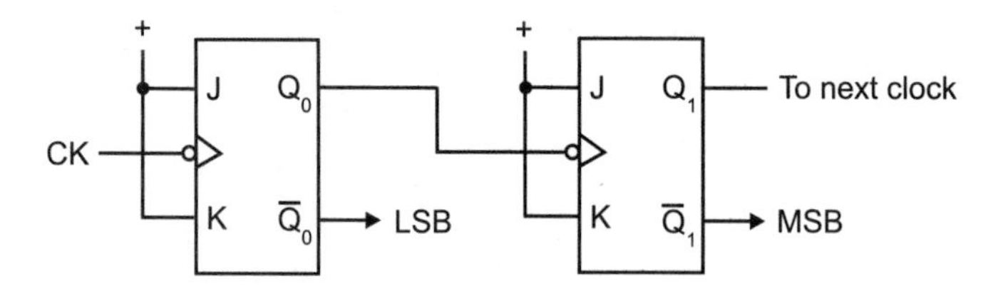


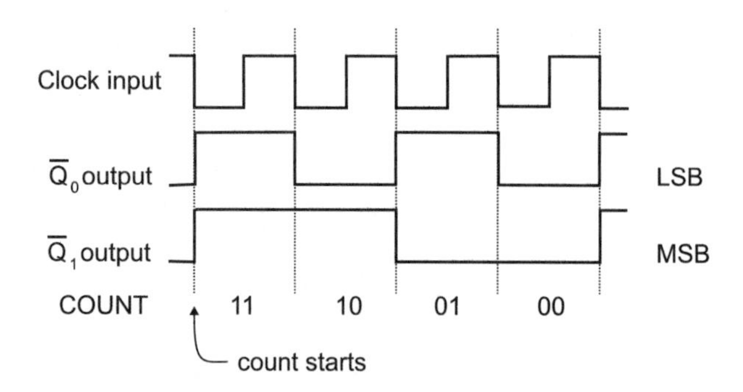


Notice that it still acts as a divide-by-four frequency divider.

## Using NOT Q for the clocks

Let's see what happens if we return to our starting point in Figure 14.2, but this time we use the NOT Q outputs to supply the clock for each IC but stay with the Q for the count. What do you think will happen?

The modified circuit is shown in Figure 14.5. The result is that, once again, it will count down.

**Figure 14.5**
Clocking from NOT Qs

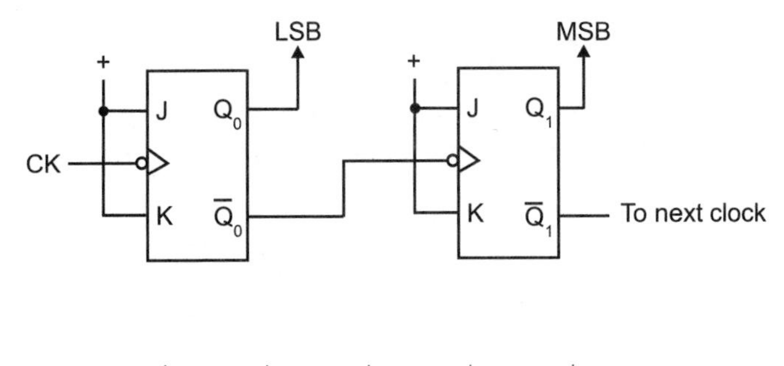


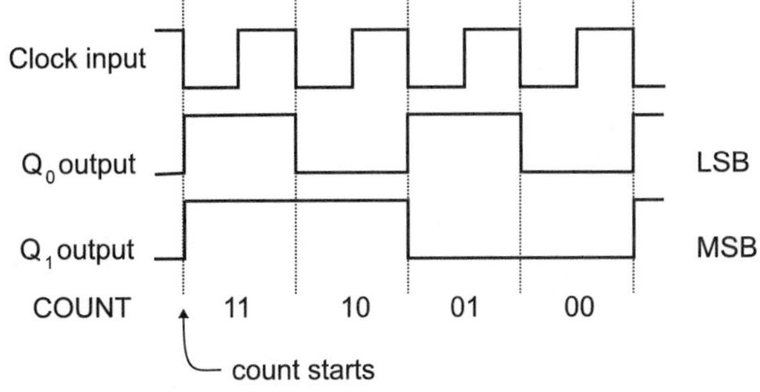


**The last possibility**

We will round this off with the last possibility: the count and the clock both coming from the NOT Q. This is the exact opposite arrangement to the one we started with in Figure 14.2.

Do you have a hunch about this outcome? Have a look at Figure 14.6 and see if you were right.

**Figure 14.6**
Count and clock from the NOT Qs

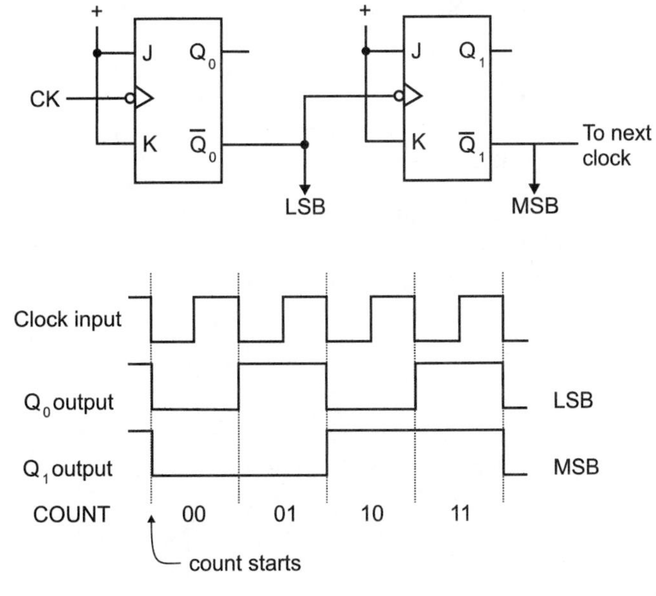

### Trailing edge summary

If we summarize the results as in Figure 14.7, we may be able to spot a pattern. Doing this will save drawing out pages of waveforms to work out what would happen with a given circuit.

**Figure 14.7** Negative-edge counters

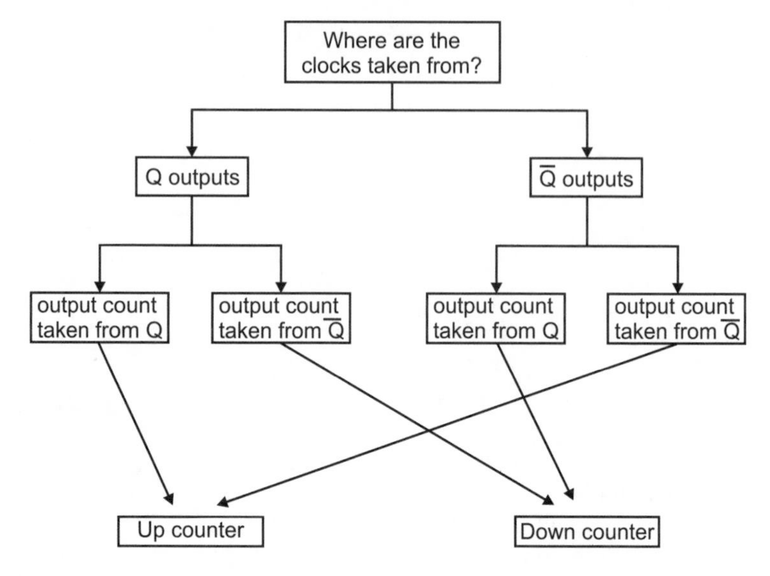


> **The golden rule is:**
>
> **In trailing edge bistables, whenever the same pin supplies the next clock *and* the count output, the result is an UP COUNTER. In all other cases it's a down counter.**

### Leading edge counters

If we use leading edge bistables we can try out a basic mod-4 counter. This is shown in Figure 14.8.

We can see from the timing diagram that this counter is a down counter. Now, this is opposite to the result that we got from the negative edge bistables under the same conditions. This is not a fluke – all the results we achieved from the negative edge triggered bistables are reversed when we come to the positive edge counters.

> **The golden rule is:**
>
> **In leading edge bistables, whenever the same pin supplies the next clock *and* the count output, the result is a DOWN COUNTER. In all other cases it's an up counter.**

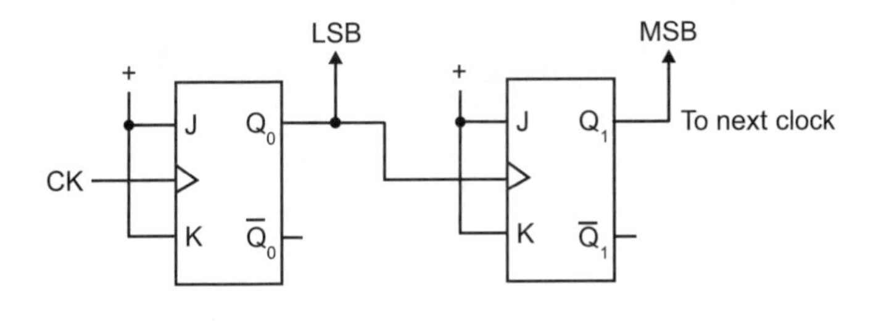


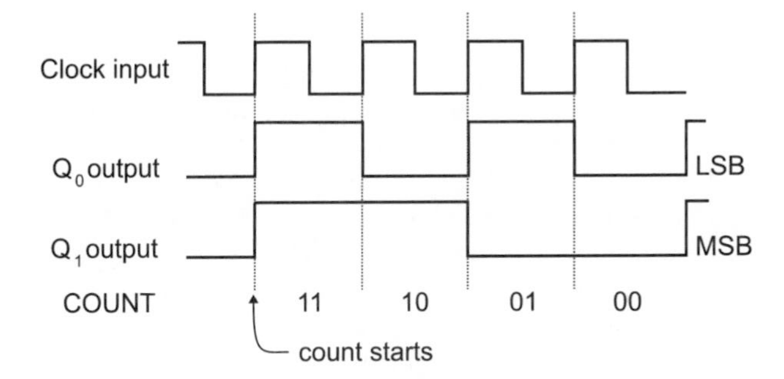


**Figure 14.8** A leading edge counter

We have seen that we can build a counter or frequency divider from positive or negative edge bistables to count up or down. But they can only count up to 2, 4, 8, 16, or any other power of 2. What about the other numbers?

## Count up to anything

This could also be entitled 'count down to anything'. Up counters have been used in the examples, but the same procedures can be followed when we are dealing with down counters.

If we had three bistables connected as an up counter, their natural count would proceed 0, 1, 2, 3, 4, 5, 6, 7 and then go back to 0 to start again.

To modify the count to be 0, 1, 2, 3, 4, 5 and then back to 0 we have to find a way to stop the count from going to 6 and instead switch the output back to 0.

Switching back to 0 is easily done by putting logic 0 onto the $\overline{CLR}$ input. Our main problem is applying this 0 at the right time. To do this, we kill off the count of 6 and replace it by a 0 output.

## The decoder circuit

When the count of 6 occurs the output from the JKs will be $Q_0 = 0$, $Q_1 = 1$ and $Q_2 = 1$, so what we need to do is to design any gate circuit that will provide an output of 0 when the three inputs are at 0, 1 and 1 and connect this 0 output to the clear inputs to each of the JKs. All three will reset and the count will start again.

Before we rush into building such a circuit we should think carefully to see if it can be done more simply by checking the sequence of the outputs.

This is:

| $Q_2$ | $Q_1$ | $Q_0$ | |
|---|---|---|---|
| 0 | 0 | 0 | |
| 0 | 0 | 1 | |
| 0 | 1 | 0 | |
| 0 | 1 | 1 | |
| 1 | 0 | 0 | |
| 1 | 0 | 1 | |
| 1 | 1 | 0 | (replace this line with 0, 0, 0) |

Looking down the columns, we have to find some combination of values that is unique to the line we want to replace. The answer, in this case, is that it is the first time when $Q_1$ and $Q_2$ are both at logic 1. So, this is enough for us. The logic circuit needed is any one that has two inputs of logic 1 and provides an output of logic 0. That's easy. A 2-input NAND gate will do that nicely; we don't need the complication of the third input.

All we do then is add the NAND gate to the circuit. Its input is taken from the $Q_1$ and $Q_2$ outputs, and its output is used to clear $Q_1$ and $Q_2$. It is often a good idea to clear all three bistables, even though $Q_0$ is already at 0. It has the advantage of ensuring that we don't forget the CLR input and accidentally leave it floating. This would run the risk of

**Figure 14.9** Modified to count 0–5

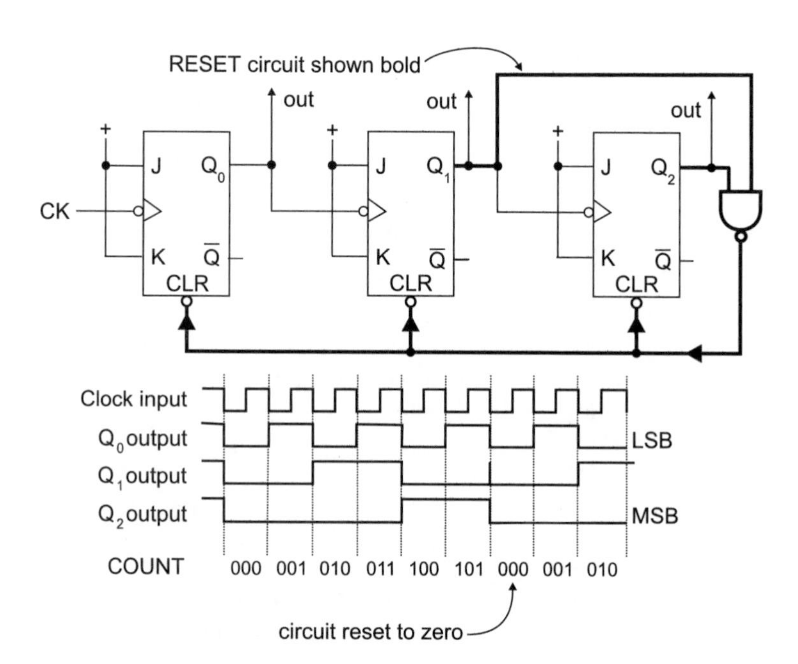

erratic performance if the floating input went low for a moment and reset this bistable. The complete circuit is shown in Figure 14.9, with the decoder circuit printed in bold.

**A small point**

If we look closely at the $Q_1$ output waveform, we can see that a small spike occurs at the moment that the circuit is reset to zero. This will be discussed at the end of this chapter.

**Example**

How could we make the circuit count from 0 to 2 and then reset?

Answer:

Looking at the count sequence:

| $Q_2$ | $Q_1$ | $Q_0$ |
|---|---|---|
| 0 | 0 | 0 |
| 0 | 0 | 1 |
| 0 | 1 | 0 |

the next step is:

0 1 1

and this is the one to use to initiate the reset action. Using the 2-input NAND gate we could simply shift its inputs to $Q_0$ and $Q_1$ (see Figure 14.10).

**Figure 14.10**
Now it counts 0–2

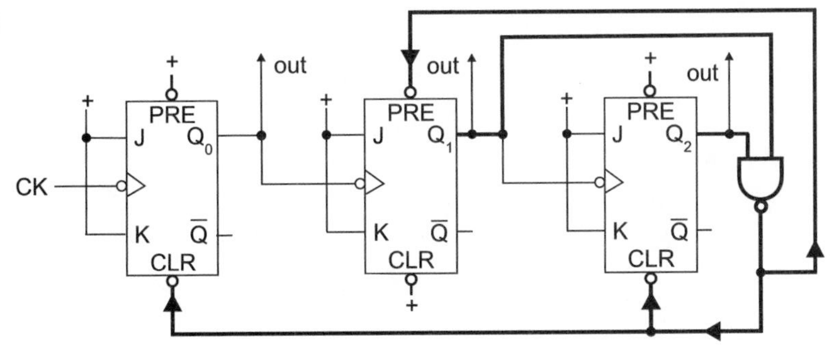


**Example**

What is the simplest way to achieve a count of 0, 1, 2, 3 and start again?

Answer:

Simply take the outputs from $Q_0$ and $Q_1$ and just ignore $Q_2$; we don't need a reset circuit. Don't forget the simple answers.

**Example**

Here is an incorrect answer. Can you spot the mistake that was made?

We wanted to count from 000 to 011 then reset.

The count went:

| $Q_2$ | $Q_1$ | $Q_0$ |
|---|---|---|
| 0 | 0 | 0 |
| 0 | 0 | 1 |
| 0 | 1 | 0 |
| 0 | 1 | 1 |

and the reset count is $Q_2 = 1$, $Q_1 = 0$ and $Q_0 = 0$.

Since we know that $Q_1$ and $Q_2$ are at logic 0 when we want to reset, we have a good idea. We use the NOT Q outputs to feed the NAND gate because they will be at logic 1 at the required time.

We build the decoder circuit in Figure 14.11. We switch it on and nothing happens, there is no output count at all.

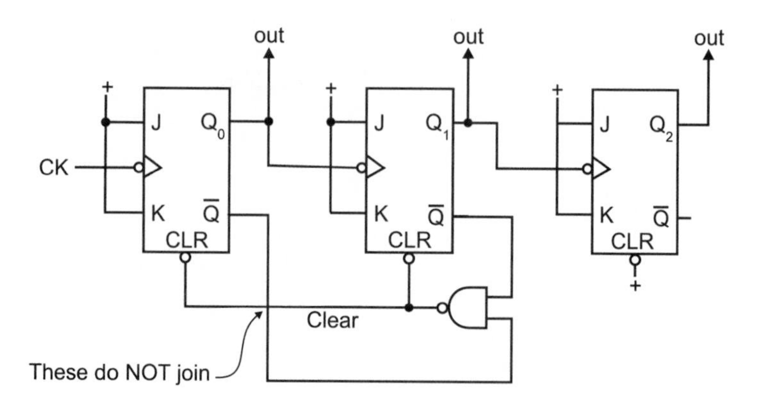


**Figure 14.11** Why doesn't this circuit work?

What did we do wrong?

Answer:

When the circuit resets, all the Q outputs go to logic 0. This means all the NOT Q outputs go to logic 1. These logic 1s are applied to the NAND gate, which has a logic 0 output, which resets the circuit.

The circuit is held in a continuous loop so it stays permanently in its reset, or clear, condition.

There is nothing inherently wrong with using a NOT Q output if it is convenient. A correct answer is shown in Figure 14.12.

**Figure 14.12**

A better answer

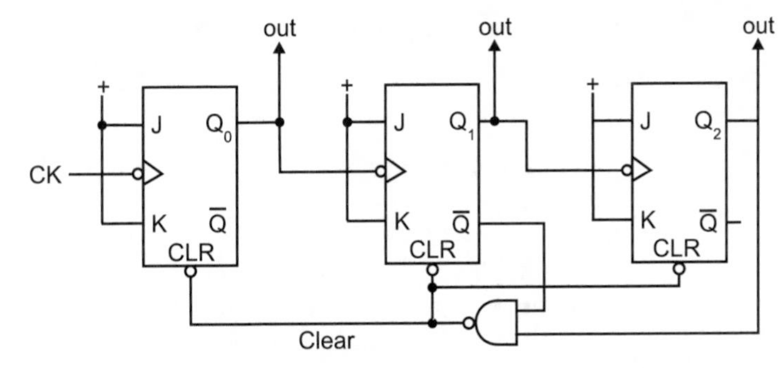


Notice that this is *a* correct answer, not *the* correct answer. We can use whatever gates we like, providing the counter works.

## Using the presets

So far, we have seen how to stop the count at any particular number – and, incidentally, how to build a frequency divider for any required ratio. In much the same way we can use the preset inputs to start the count at any particular value.

The only change is to take the output back to preset or clear as necessary.

In Figure 14.9, we stopped counting at a count of 101 and then returned to 000 to start again. Now we will modify the circuit to count from 010 to 101.

The output count will be 010, 011, 100, 101 and then 010.

The only change will be to move the connection going to the CLR input of the second bistable up to the preset input to switch $Q_1$, as in Figure 14.13.

**Figure 14.13**

Starting a count at 010

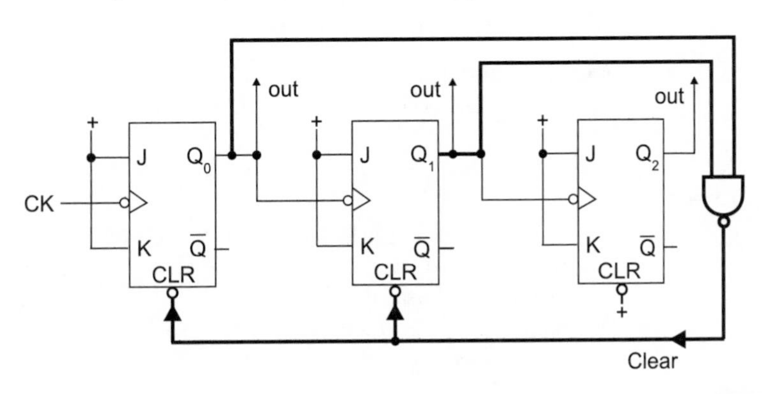

## Why are these called 'asynchronous' counters?

As well as asynchronous, they are also called 'ripple' counters or 'ripple-through' counters.

Ripple-through is quite a good name because it describes the action nicely. The point here is that the clock is only applied to the first bistable; it switches and only then does it provide the clock to the second bistable. The clock to the second bistable has been delayed by the propagation time of the first bistable. When we have a large counter, there is a significant delay as the signal 'ripples through' the various stages.

The clock is therefore not applied at exactly the same moment to all the bistables – hence they are not perfectly synchronized. For this reason they are called asynchronous counters. More about this in a moment.

There are three problems with asynchronous counters;

1 The race problem.
2 Glitches.
3 'Clock skew' or 'dynamic hazard'.

### The race problem

This problem occurs at the moment when the count is reset by the voltage being fed to the clear inputs.

The decoded output tries to reset all the counters to zero but, and here's the problem, as soon as one of them does reset, the decode circuit is switched off.

If we glance back to Figure 14.9 we reset the circuit as soon as the output of $Q_1$ and $Q_2$ have changed to logic 1. Now let's assume that, due to manufacturing tolerances or different logic families, the middle bistable is very fast and the one on the right is very slow. As soon as the NAND output applies logic 0 to the CLR inputs, the middle bistable will rapidly reset. The NAND gate will notice that one of its inputs has gone low and will change its output to logic 1. This cancels the CLR command to the bistables. If the slow bistable on the right still hasn't changed, then it may not do so.

This means that the count may not restart from zero. We have a race. Can all the bistables be cleared (or set) before the control voltage is removed?

### Glitches

In the timing diagram in Figure 14.9 we can see that a small spike occurs on the $Q_1$ output just as the circuit resets. The spike occurs because all the bistables used for decoding must be at logic 1 before the NAND gate can reset the circuit. However fast the NAND gate is, the spike must occur first.

This 'glitch' is very difficult to detect by test instruments, and is bad news in digital circuits. Glitches can cause all manner of erratic behaviour and odd symptoms. If we are using an asynchronous counter in a slow circuit, such as controlling light displays, the spike exists for such a short time that we cannot detect it but the logic circuit may still malfunction for no obvious reason.

### 'Clock skew' or 'dynamic hazard'

This is another trick digital circuits can play to give us hours of fault-finding fun.

In Figure 14.14 we have the output from a counter as it counts up from 000 to 100 (or 0 to 4 in denary).

In reality, as we found with glitches, the JKs will take time to respond. Their propagation time will mean that after $Q_0$ switches, a few nanoseconds will pass before $Q_1$ responds and then a further few nanoseconds as we wait around for $Q_2$ to provide its output.

This means that the timing diagram in Figure 14.14 is not telling the whole truth. The edges of all waveforms are shown aligned, whereas we know that they are not.

**Figure 14.14**
A straightforward count – or is it?

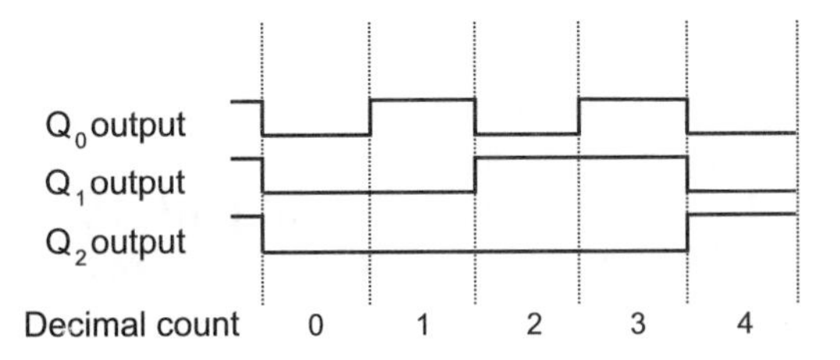


In Figure 14.15, we expand the time scale enormously until we can see the propagation time. Notice that the trailing edges, which were previously shown in vertical alignment, are now shown at an angle,

**Figure 14.15**
What a count!

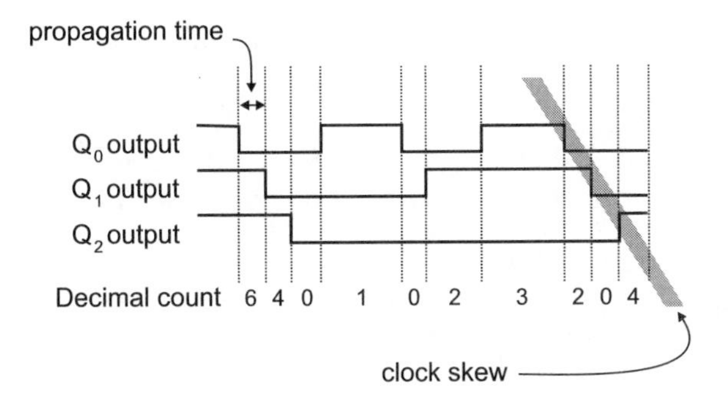

giving rise to the term 'clock skew'. More important than this – we get a real shock when we look at the count.

These time delays, as we have seen, have the effect of throwing up spurious counts. Clock skew or dynamic hazard is a real problem in digital circuits, and sometimes accounts for 'inexplicable' failures in circuits. It is even more puzzling when the output is displayed on a light panel since the lights (and our eyes) tend to respond only to the main count and hide the true count, which is playing havoc with the rest of the (faster) digital circuit.

Remember too that the timing diagram in Figure 14.15 is necessarily distorted. The real propagation time may be 5 ns per bistable. As drawn the delay is about a millionth of the width of the clock pulse, and therefore if it was drawn to scale the skew would not be apparent. Even using instruments like oscilloscopes the skew is often missed, and the circuit is said to have 'decided' not to work.

Like glitches, these failures are very difficult to detect, so we need to watch out for them.

## Quiz time 14

In each case, choose the best option.

**1 How many T-type bistables would be needed to divide a clock frequency by 64?**

(a) 32
(b) 100 000
(c) 6
(d) 5

**2 Dynamic hazard is caused by:**

(a) the preset inputs being connected to logic 1 levels.
(b) D-type bistables.
(c) propagation delay in the bistables.
(d) decoder circuits.

**3 If a small circle is shown on the input of a bistable, this indicates that the input is:**

(a) a reset pin.
(b) active low.
(c) not to be used.
(d) active high.

**4 If an up counter uses negative edge bistables and takes its output from the NOT Q outputs, the clock for each bistable is taken from:**

(a) the previous Q output.
(b) the previous clock.
(c) either the previous Q or NOT Q.
(d) the previous NOT Q output.

**5 A decoding circuit is used to change the:**

(a) maximum and minimum value of the count.
(b) speed of the count.
(c) amount of skew.
(d) direction of the count.

# 15

# Synchronous and integrated counters

## Synchronous counters

### Advantages

Both race problems and dynamic hazards can be overcome by using synchronous counters, also called 'parallel' counters.

The main difference in the design of synchronous counters is that the original clock pulses are taken to all bistables, not just the first. The clock is therefore fed in parallel to all bistables – hence the alternative name of 'parallel' counter.

This has the effect of making the synchronous counter faster because all bistables switch at the same time, so we don't have to wait for the change to ripple through the chain of bistables.

### Disadvantages

We have mentioned in Chapter 11 that gates cause a momentary increase in current at the moment of switching. Bistables cause the same problem. Synchronous counters are designed to switch the bistables at the same moment, so the accumulated current burst can be severe. This requires power supplies with higher current capability and much improved decoupling.

The only other small disadvantage is that the circuitry is very slightly more complex. But only slightly; not enough to worry us.

## A synchronous up counter.

Note in Figure 15.1 the three significant features:

1 The clock goes directly to each bistable
2 There is extra decoding compared with the equivalent asynchronous counter
3 Only the J and K inputs to the first bistable are held to logic 1.

**Figure 15.1**
A synchronous up-counter

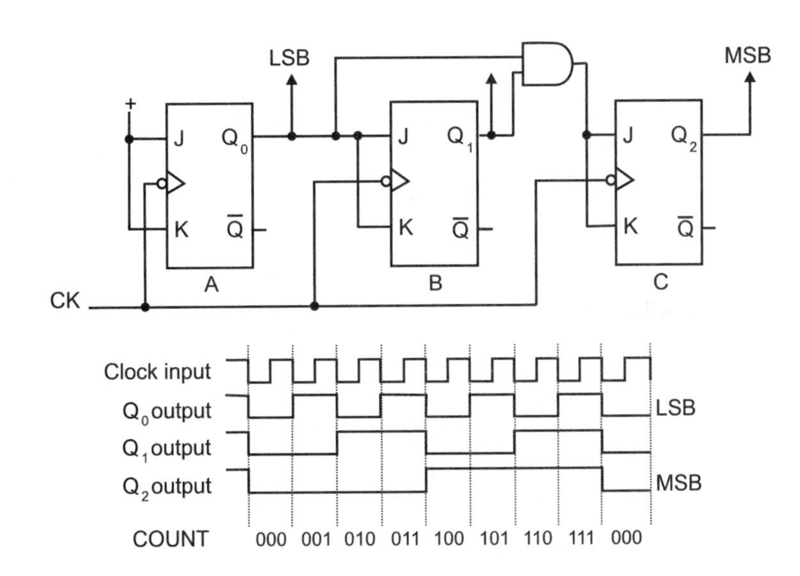


### Action of the circuit

The bistables have been labelled A, B and C, and let's assume that at the start all the outputs are at logic zero so $Q_A = Q_B = Q_C = 0$. The bistables are master–slave JKs.

The AND gate is there to detect the moment when the output of the first two bistables is at logic 1. This detects a binary count of 11 and sends a signal to switch on the next bistable to give the next count as $100_2$.

### First clock pulse

State of inputs before the clock pulse: J, K of bistable A = 1; J, K of bistable B = 0; J, K of bistable C = 0.

Action: when the trailing edge of the clock occurs, bistable A toggles but B and C remain at zero because their J and K inputs were held at logic 0.

Output state after the clock pulse: $Q_A = 1$, $Q_B = 0$, $Q_C = 0$, giving a binary count of 001 or 1 in denary.

### Second clock pulse

State of inputs before the clock pulse: J, K of bistable A = 1; J, K of bistable B = 1; J, K of bistable C = 0.

Action: Bistables A and B will toggle but bistable C will remain at logic 0.

Output state after the clock pulse: $Q_A = 0$, $Q_B = 1$, $Q_C = 0$, giving a binary count of 010 or 2 in denary.

### Third clock pulse

State of inputs before the clock pulse: J, K of bistable A = 1; J, K of bistable B = 0; J, K of bistable C = 0.

Action: Bistable A toggles, bistable B remains at logic 1 and bistable C remains at logic 0.

Output state after the clock pulse: $Q_A = 1$, $Q_B = 1$, $Q_C = 0$, giving a binary count of 011 or 3 in denary.

### Fourth clock pulse

State of inputs before the clock pulse: J, K of bistable A = 1; J, K of bistable B = 1; J, K of bistable C = 1 (AND gate now activated).

Action: All bistables will toggle.

Output state after the clock pulse: $Q_A = 0$, $Q_B = 0$, $Q_C = 1$, giving a binary count of 100 or 4 in denary.

The output waveforms are exactly the same as those we saw with the asynchronous counter except for the absence of clock skew.

### Adding more bistables for a higher count

We can carry on adding more bistables to give any count that we require. If we wish to add further bistables, each would be fed via another 2-input AND gate with its input fed from the previous two bistables. It is tempting to think that the next JK would need a 3-input AND gate to recognize the time when $Q_A$, $Q_B$ and $Q_C$ are all at logic 1. However, the existing AND gate provides us with A.B, so by adding another 2-input AND gate and a little Boolean we achieve (A.B).(C) = A.B.C. An extended counter is shown in Figure 15.2.

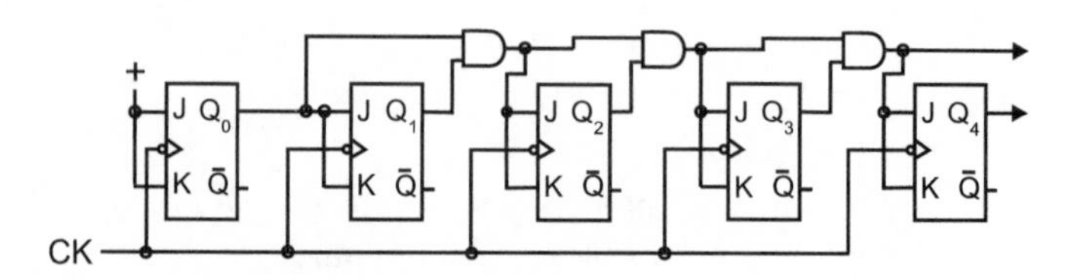


**Figure 15.2** Synchronous counters can be easily extended

### Up counter or down counter?

If we are using negative edge triggering, look to see where the output is taken from. If it's the same point that is being used to provide the clock input to the next bistable then we have an up counter; if not, it's a down counter.

With a positive edge triggered circuit, ask the same questions but expect the opposite answers.

### Modulo-n counters

Counting up to any number, or between two numbers, is achieved by adding decoding circuits using the clear and preset inputs as we did with asynchronous counters. In fact, the circuitry is exactly the same.

Remember, though, that this decoder circuitry to control the start and stop points of the count is in addition to the AND gates that we added to build the basic synchronous counter, so the circuit can look pretty impressive by the time it is complete.

### Example

Figure 15.3 shows a counter. Study the diagram and answer the following questions:

**Figure 15.3**

So, what have we got here?

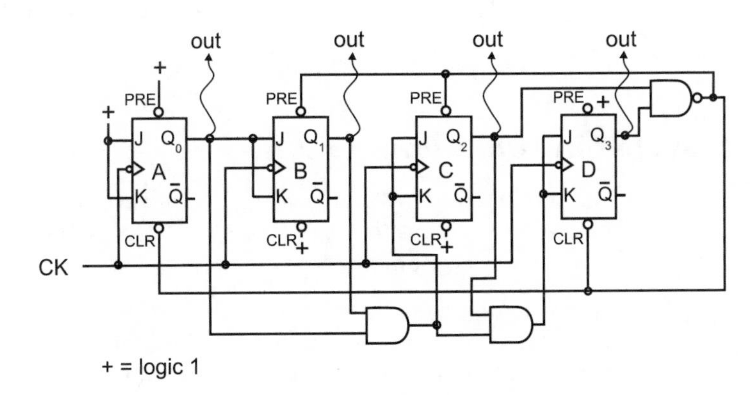


1 Is this a parallel counter or a ripple counter?
2 Is it an up counter or a down counter?
3 At what binary value does the count start?
4 What is the last count value (in binary)?
5 What is its mod value?
6 If the outputs were moved from the Q to the NOT Q outputs, how would the count proceed? (Note: no other changes are made.)

Answers:

1 Check to see whether the same clock input is connected to all bistables. It is, so it's a parallel counter.

2 It's an up counter. The outputs are taken from the same points as the J and K inputs to the next bistable.
3 When the decoding circuit produces a low, bistables B and C are preset and A and D are cleared. The output is therefore $Q_A = 0$, $Q_B = 1$, $Q_C = 1$ and $Q_D = 0$. This gives a starting count of 0110 in binary, 6 in denary.
4 The reset NAND gate operates the first time bistable C and D are both set. This will occur at a count of 1100 in binary, 12 in denary. Now remember, this count is killed off by the reset circuit and so the last count is the one immediately before this one – that is, binary 1011 or denary 11.
5 Mod-6. The count is 6, 7, 8, 9, 10, 11 and then 6 again. This gives six different counts, which is mod-6.
6 In denary it would count down: 9, 8, 7, 6, 5, 4 and then reset to 9.

Here's how to work out the final answer.

The limits of the count when using the Q outputs were 6 and 11. When it was counting six the output binary values would be: $Q_A = 0$, $Q_B = 1$, $Q_C = 1$ and $Q_D = 0$. The NOT Q values are just the opposite, so they would be NOT $Q_A = 1$, NOT $Q_B = 0$, NOT $Q_C = 0$ and NOT $Q_D = 1$. The value is 1001 in binary or 9 in denary.

The other end of the count would be denary 11 or binary $Q_A = 1$, $Q_B = 1$, $Q_C = 0$ and $Q_D = 1$. Inverting each bit will give us NOT $Q_A = 0$, NOT $Q_B = 0$, NOT $Q_C = 1$ and NOT $Q_D = 0$. This is binary 0100 or denary 4.

## Integrated bistables and counters

As mentioned in the previous chapter, integrated counters are usually easier to use than stringing a series of bistables together. They win on cost, power consumption, ease of use and space.

A couple of general points. Not all bistables are available in all logic families. Those chosen here are readily available in the HC family – which, at the moment, is our first choice for new designs. All of these ICs are also available in the traditional 74LS family.

## 74HC74, 74LS74

This is a dual D-type positive edge triggered flip-flop with clear and preset.

The 'dual' just means we have two of them in the same package. They are quite independent, and if only one is required just ignore the other one.

The pin-out diagram is shown in Figure 15.4 and the truth table is in Table 15.1. Its maximum switching frequency is about 25 MHz.

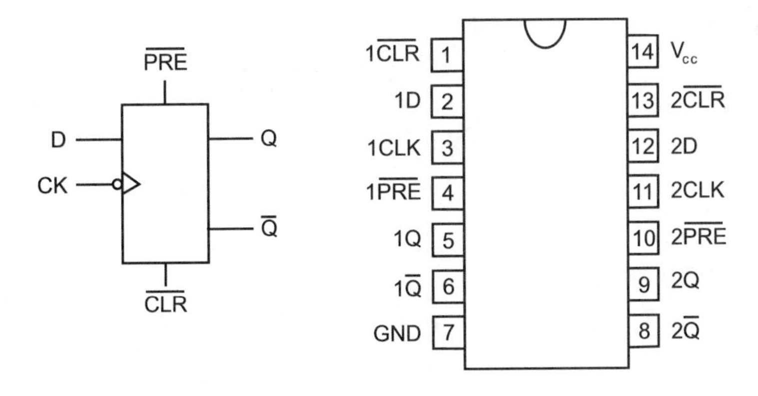


**Figure 15.4** The 74XX74 D-type

**Table 15.1** Truth table of the 74XX74

| *Inputs* | | | | *Outputs* | |
|---|---|---|---|---|---|
| $\overline{PRE}$ | $\overline{CLR}$ | *CLK* | *D* | *Q* | $\overline{Q}$ |
| L | H | X | X | H | L |
| H | L | X | X | L | H |
| L | L | X | X | H | H |
| H | H | ↑ | H | H | L |
| H | H | ↑ | L | L | H |
| H | H | L | X | $Q_0$ | $Q_0$ |

**Truth table symbols**

X means 'don't care'; the input can be held at either logic level or any voltage in between.

H, L refers to the logic levels. H is high and refers to logic 1, L is low and equates to logic 0. This is the normal way of showing the voltage levels in truth tables used to describe the operation of a device. It can also be found when describing the operation of logic gates.

↑ refers to the leading or positive-going edge of the pulse.

$Q_0$ refers to the state of Q before the current clock pulse.

Note how it is possible to take both the preset and the clear inputs to a low state and force both Q and NOT Q to a high state. This is an unstable state, and it doesn't last once either or both of them return to their normal high state.

### The truth table, line by line

The first three lines of the truth table just show that the preset and clear inputs control the output regardless of the other inputs.

The last line shows that, if the clock is missing, everything stops.

The remaining lines show what we regard as the normal operation of the D-type bistable. An input is applied to D, a clock pulse is applied and the D input state is copied to the output.

Figure 15.5 shows the connections necessary to make the D-type operate correctly.

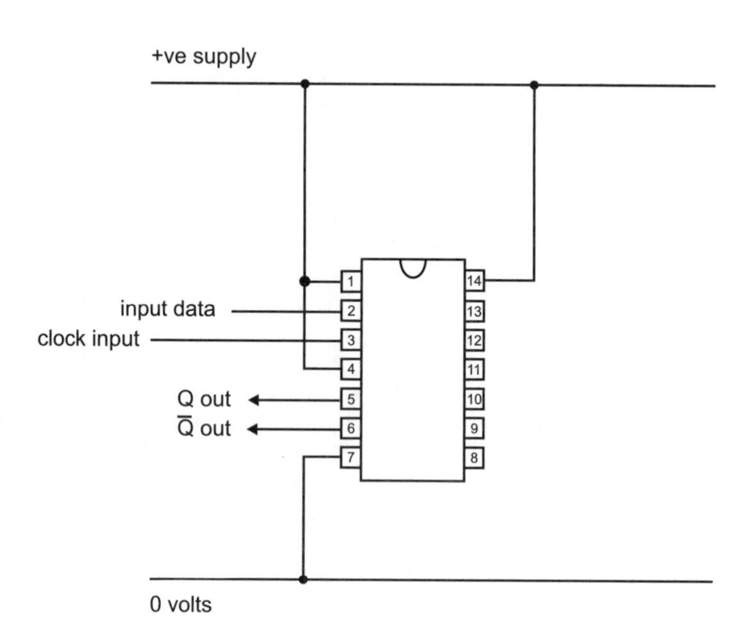


**Figure 15.5** One of the D-types all wired up and ready to go

## 74HC112, 74LS112

This is a dual master–slave JK with data transfer occurring on the trailing edge.

The active low preset and clear inputs are disabled during the set-up time that extends for the 25 ns or so immediately before the data transfer. During this set up time it reads the input data from J and K. Its maximum switching frequency is about 25 MHz, similar to the 74XX74.

The pin-out diagram is shown in Figure 15.6 and the truth table is in Table 15.2.

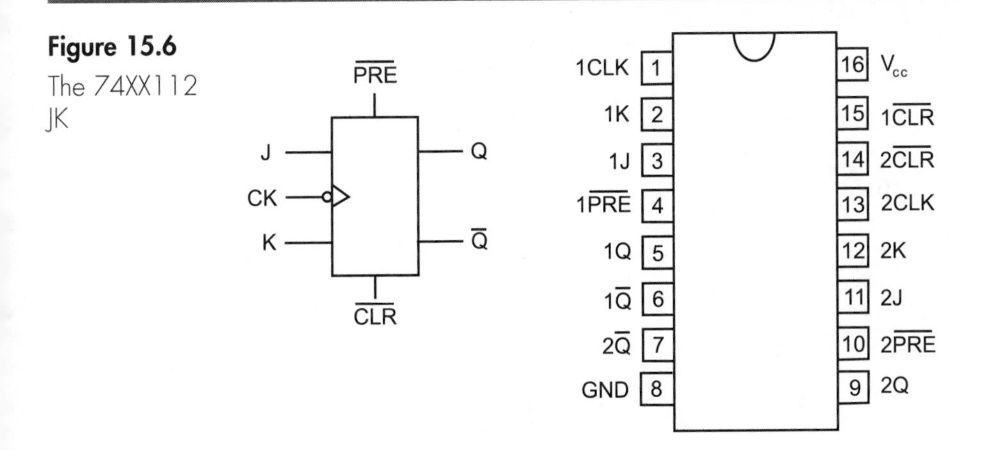


**Figure 15.6** The 74XX112 JK

**Table 15.2** Truth table of the 74XX112

| *Inputs* | | | | | *Outputs* | |
|---|---|---|---|---|---|---|
| $\overline{PRE}$ | $\overline{CLR}$ | *CLK* | *J* | *K* | *Q* | $\overline{Q}$ |
| L | H | X | X | X | H | L |
| H | L | X | X | X | L | H |
| H | L | X | X | X | H | H |
| H | H | ↓ | L | L | $Q_0$ | $\overline{Q_0}$ |
| H | H | ↓ | H | L | H | L |
| H | H | ↓ | L | H | L | H |
| H | H | ↓ | H | H | Toggle | Toggle |
| H | H | H | X | X | $Q_0$ | $\overline{Q_0}$ |

Apart from the obvious fact that it requires both J and K inputs as opposed to the single D input of the 74XX74, there are no surprises at all. To accommodate the J and K inputs, the pin-out has been extended to a 16-pin package.

## The truth table, line by line

Most of this follows the pattern used in the 74XX74 truth table.

The first two lines show the effect of the asynchronous preset or clear being taken low. The third line refers to the unstable state that occurs only while the preset and clear inputs are both held low. Notice again that these inputs over-ride all the other inputs.

The next four lines show that changes occur on the trailing edge of the clock and they follow the normal responses to change in the level of J and K. The seventh line shows the toggle state.

Finally, the last line shows the obvious nil response to no clock input.

Figure 15.7 shows the connections for a mod-4 synchronous up counter.

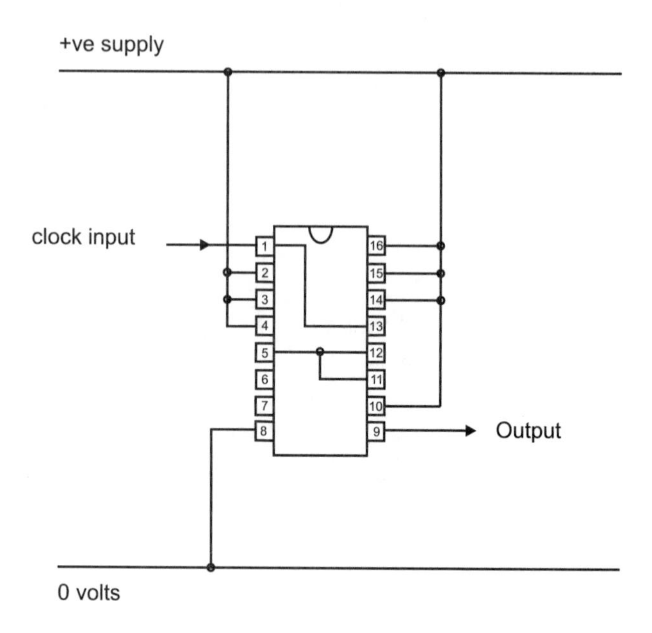


**Figure 15.7**
A JK synchronous modulo-4 counter

## 74XX193

This is a 4-bit synchronous up/down counter with asynchronous clear.

As the name would suggest, there are four bistables inside connected as a synchronous counter. The pin-out diagram is shown in Figure 15.8.

It has two separate input clocks; one makes it count up on each positive edge of the waveform and the other causes it count down. Either clock can be used at any time provided the other clock is held at a high logic state, so we can mix up and down counting anywhere from 0–15 in binary. As with most other counters there is no minimum operating frequency, so it can be used for slow inputs such as people being counted through a turnstile.

The clear input, sometimes called the master reset, is active high and will over-ride all other inputs. This means that it must be held at logic low to allow the count to proceed. This reset is asynchronous. This means that it will clear the input immediately. Some counters have synchronous clears or resets that take effect on the next clock pulse.

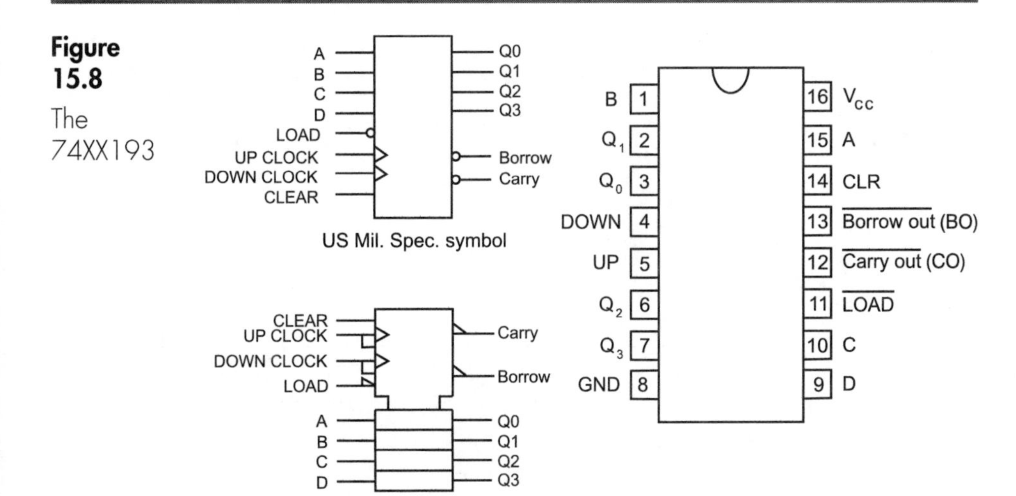


**Figure 15.8** The 74XX193

We can join up these bistables to form larger counts by connecting the pin BO to the next chip's 'down clock' and the CO pin to the other chip's 'up clock'. For larger counts there are other chips that are easily available, like the 74HC4060 which is a 14-bit counter.

## Operation

Refer to Figure 15.9.

Let's assume we would like to load it with the number 7 then count up from 8 to 12 and stop.

Step 1: We start by applying a positive-going pulse to clear the outputs $Q_0$ to $Q_3$. This makes all the inputs to return to zero regardless of all other inputs.

Step 2: Next, we enter the starting values – in our case A = 1, B = 2, C = 3 and D = 0. When all is ready, we apply a negative-going 'load' pulse. The timing of the load pulse is not critical; it can be delayed as long as we like. This can be the 5 minutes while we stare at the vending machine wondering whether to have coffee or tea. As the load pulse occurs, the outputs $Q_0$–$Q_3$ are set to the loaded values so our outputs go to $Q_0 = 1$, $Q_1 = 1$, $Q_2 = 1$ and $Q_3 = 0$.

Step 3: Nothing happens now until a clock pulse occurs. As mentioned a moment ago, there are two clocks, one for counting up and one for counting down. The unused clock must be held at logic high. We wanted to count up from 8 to 12, so we need five positive edges to occur.

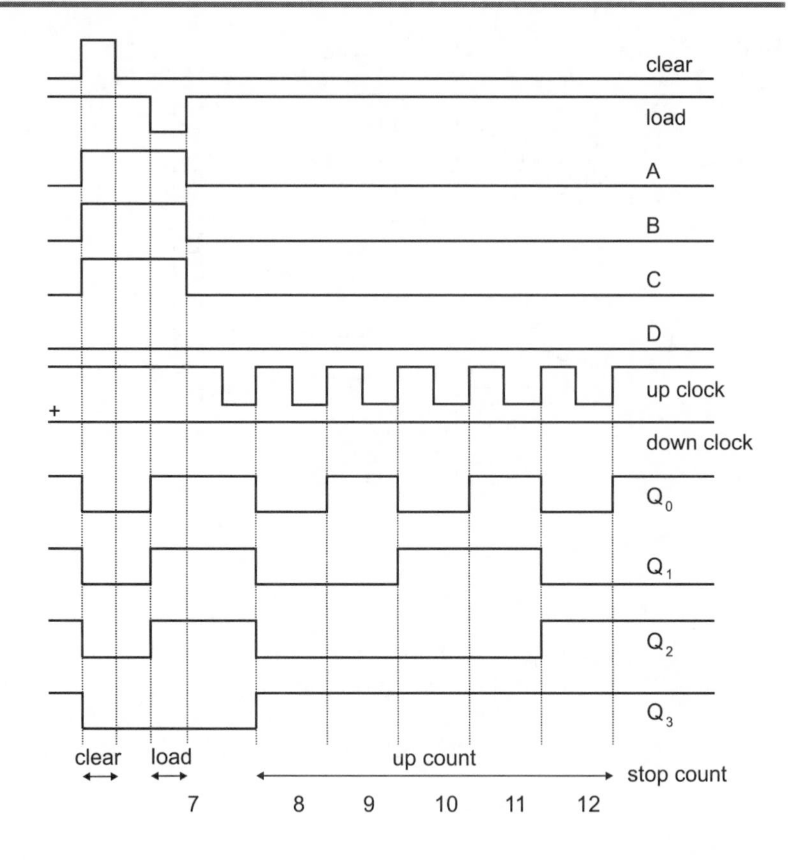


**Figure 15.9** Timing diagram of a 74XX193

Step 4: To stop the count, we just hold both clock inputs at a high level. If we wished, we could now continue counting or apply a clock pulse to the 'down clock' to count downwards.

### Good value for money

Inside this IC we get five bistables and 47 gates in a ready-built circuit, all for the cost of two bistable chips! This is a really nice one to use for counting experiments.

## Decade counters

Since we live in a denary world, it is often the case that we wish to count up to 10. We can achieve this by designing any form of counter to reset after 10 counts. Then by connecting a series of them together we can easily extend the counter to 100 or 1000, or whatever we want. This is ideal if we are counting events like products on a conveyor belt or litres of petrol being pumped into a car.

For convenience, some counters are available that are ready-made for a count of 10 and are, of course, referred to as denary counters. An example is the 74XX390.

## 74XX390

This is a dual version of the 74XX90. The benefit of the dual version is that these can be easily connected to provide a single chip count up to 100.

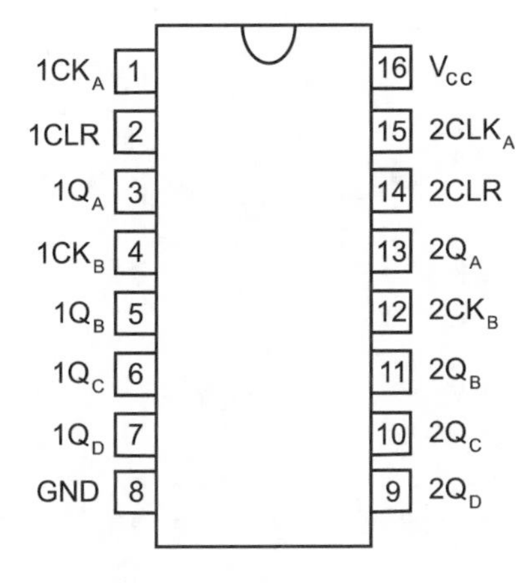


**Figure 15.10**
The 74XX390

It is sometimes referred to by the scary title of a bi-quinary counter. 'Quinary' means a count of five, and the 'bi' just means that we can count the five twice. Two fives are 10. It's now a not-so-scary counter. The advantage of making it in a bi-quinary form is that, by suitable choice of interconnections, we can achieve a count of many multiples of two, five and 10.

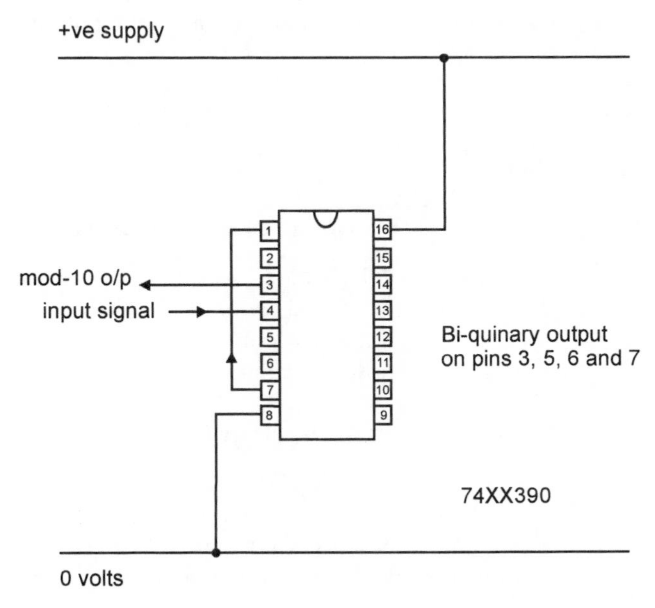


**Figure 15.11**
The 74XX390 as a bi-quinary decade counter

**Table 15.3** The bi-quinary count using the 74XX390

| *Count* | *Outputs* | | | |
|---|---|---|---|---|
| | $Q_A$ | $Q_D$ | $Q_C$ | $Q_B$ |
| 0 | L | L | L | L |
| 1 | L | L | L | H |
| 2 | L | L | H | L |
| 3 | L | L | H | H |
| 4 | L | H | L | L |
| 5 | H | L | L | L |
| 6 | H | L | L | H |
| 7 | H | L | H | L |
| 8 | H | L | H | H |
| 9 | H | H | L | L |

Note: The Q outputs are in an unusual order. The first bistable provides the 'bi' part of the count at $Q_A$, then $Q_D$, $Q_C$ and $Q_B$ count up to five

Figure 15.10 shows the chip pin-out. The two-decade counters are indicated by the '1' or '2' at the start of each pin designation. The mod-2 input is called clock A or $CLK_A$, and its output is at $Q_A$. The mod-5 part uses the input $CLK_B$, and the outputs are on $Q_B$, $Q_C$ and $Q_D$. Each decade counter has its own clear.

Figure 15.11 shows the chip connected as a decade counter. The output count is shown in Table 15.3.

For a mod-100 counter, just connect the output from $Q_A$ to input B of the next counter.

## Quiz time 15

In each case, choose the best option.

**1 A synchronous counter is also called:**

(a) a parallel counter.
(b) an asynchronous counter.
(c) a ripple-through counter.
(d) an up/down counter.

**2 If a synchronous clear is activated, the outputs will be:**

(a) reset immediately.
(b) set to logic high when the clock pulse is applied.
(c) held at their current state.
(d) reset at the time of the next clock pulse.

**3 In a truth table, the symbol X means:**

(a) this facility is not available.
(b) the connecting wires should be twisted.
(c) that the logic state is irrelevant.
(d) no logic level can be applied.

**4 The term $Q_{n+1}$ most likely means the logic state after:**

(a) $Q_{n+2}$
(b) $Q_{n-1}$
(c) $Q_A$
(d) $Q_n$

**5 A synchronous counter:**

(a) must use trailing edge JKs.
(b) is faster than an asynchronous counter.
(c) is simpler to build than an asynchronous counter.
(d) causes clock skew.

# 16

# Some more counters, codes and registers

## The ring counter

This is an unusual counter in that it does not provide a binary count at the output but it does provide a moving sequence of outputs and a frequency divider output.

These counters are usually made from D-type bistables, although JKs can be used. They are very simple, with any number of bistables connected in a row and the final output connected back to the input to form a complete ring – hence the name.

It's an interesting counter to play with, but it is uneconomic for larger counts for two reasons. First, by its nature it only produces a single count for each bistable whereas with all the previous designs the count number has doubled for each bistable added. Secondly, when we have more than a few bistables it becomes more economic to use a purpose-built integrated circuit.

The operation can be followed in Figure 16.1. Assume that we start by applying an active low clear signal to reset the whole counter to give zero outputs. We can choose which preset inputs to activate. In this example, we set just the first bistable to keep things really simple.

### Starting point

The first bistable (A) has a high state at its output that is ready to be loaded into bistable B as soon as the clock pulse occurs. At the same time, there is a low state applied to the first bistable that is being

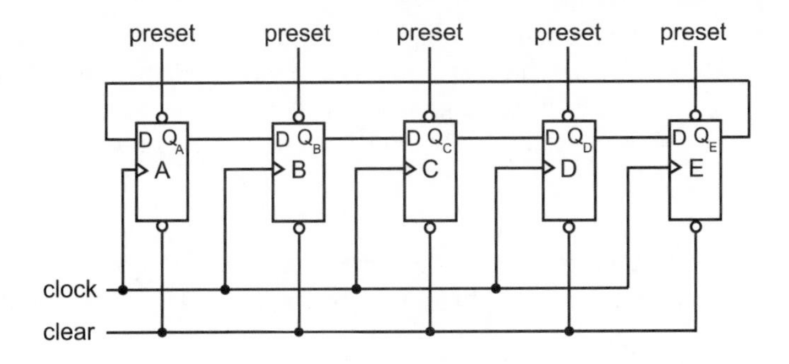


**Figure 16.1**
A ring counter or rotate register

wrapped around from the last bistable (E), again, not loaded until the clock pulse occurs. Bistables C, D and E all have logic low being loaded from the preceding bistables.

After clock pulse 1 the high level is loaded into bistable B and low levels are loaded into A, C, D and E.

After clock pulse 2 the high state is passed on to bistable C and all the others are loaded with low levels. These levels are applied to the next bistable in each case ready for the next clock pulse to be applied.

Every clock pulse has the effect of transferring the data one step to the right, and it will eventually be fed back to the start as in Table 16.1.

**Table 16.1** Ring counter or rotate register operation

| *After clock pulse* | $Q_0$ | $Q_1$ | $Q_2$ | $Q_3$ | $Q_4$ |
|---|---|---|---|---|---|
| 0 (start) | **H** | L | L | L | L |
| 1 | L | **H** | L | L | L |
| 2 | L | L | **H** | L | L |
| 3 | L | L | L | **H** | L |
| 4 | L | L | L | L | **H** |
| 5 | **H** | L | L | L | L |
| 6 | L | **H** | L | L | L |

We can take the final output from each of the Q outputs. In this way we could use this ring counter to switch five external circuits in rotation. If these external circuits control the clock pulse, then quite complex industrial processes can be performed one step at a time. This circuit is also referred to as a sequencer since it can control a sequence of tasks.

If we decide to take the output signal from just one of the bistables, then we have a frequency divider or a way of changing the mark–space ratio of the output. The mark–space ratio is the on–off ratio and can be changed by presetting a different number of bistables before the circuit is clocked.

If JKs are used, the final Q is returned to the first J input and the final NOT Q is returned to the first K input. In each of the other cases, the Q is connected to the next J and the NOT Q is connected to the next K.

## Twisted ring or Johnson counter

This is a simple modification to the ring counter which generates a different pattern much loved for light displays and advertising as it appears as a moving image. The changes appear slight. The final output from the last bistable is inverted by taking it from the NOT Q output rather than from the Q as in the ring counter, and all the bistables are cleared to start. The twisted ring is shown in Figure 16.2.

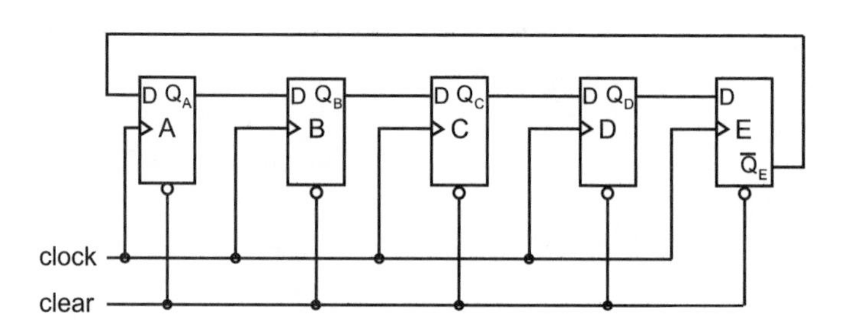


**Figure 16.2**
A twisted ring or Johnson counter

When all the bistables are cleared, the final NOT Q is at a high state that is ready to be fed into bistable A.

After the first clock pulse $Q_A$ is high, and remember that the input to the first bistable remains high because there will still be another high state coming from the NOT Q of the last bistable. As the clock pulses occur, the bistables will fill with high outputs until bistable E sets. When this occurs its Q goes high but its NOT Q output goes low, and the low state will start to be fed along the line of bistables.

Notice how the count pattern now occurs over 10 clock pulses rather than the five we had with the normal ring counter. The logic sequence is shown in Table 16.2.

**Table 16.2** Twisted ring or Johnson counter

| *After clock pulse* | $Q_0$ | $Q_1$ | $Q_2$ | $Q_3$ | $Q_4$ |
|---|---|---|---|---|---|
| 0 (start) | L | L | L | L | L |
| 1 | **H** | L | L | L | L |
| 2 | **H** | **H** | L | L | L |
| 3 | **H** | **H** | **H** | L | L |
| 4 | **H** | **H** | **H** | **H** | L |
| 5 | **H** | **H** | **H** | **H** | **H** |
| 6 | L | **H** | **H** | **H** | **H** |
| 7 | L | L | **H** | **H** | **H** |
| 8 | L | L | L | **H** | **H** |
| 9 | L | L | L | L | **H** |
| 10 | L | L | L | L | L |
| 11 | **H** | L | L | L | L |

. . . and the pattern starts again . . .

## Registers

A register is just a collection of flip-flops that are used to store or manipulate data in the form of logic 0 and logic 1 levels. With registers it is usually easier to use the alternative form of 1 and 0 rather than high and low levels.

A flip-flop can only store one bit by being set or cleared, so to handle 8 bits at a time we would need eight flip-flops and would refer to this as an 8-bit register. To save space, an 8-bit register would be shown as in Figure 16.3.

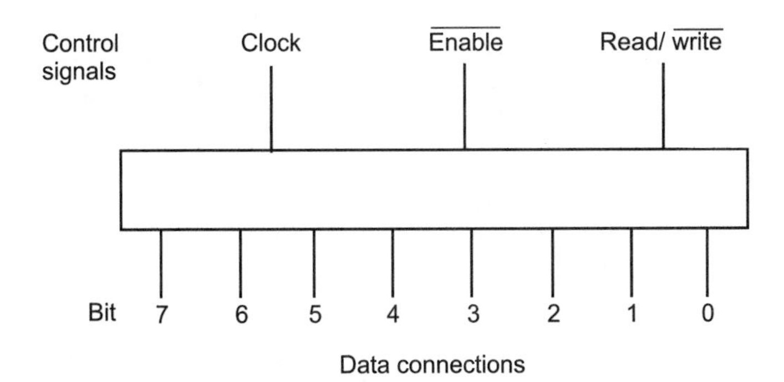


**Figure 16.3** An 8-bit register

Note how we number the bits starting at bit 0 rather than refer to the bistable outputs as $Q_A$ or $Q_0$ etc. This is because registers are often used to store data in the form of binary numbers.

The register has two distinct groups of connections: the data bits 0 to 7 and the control signals.

The data connections or data lines carry the binary levels in or out of the register, to set or clear each bistable separately. This data can be entered by switches or be derived from external circuits. The number of data lines determines the size of the register, so a 64-bit register would have 64 data connections.

Apart from the clock input, there are two new control signals:

1 Enable. This is a simple on/off switch for the register. The line over the top of the word indicates that it is active low and the register is 'on' when this line is 'low' or at logic zero. Therefore, it follows that the register is disabled or switched 'off' when the enable line is at logic 1 or 'high'. Nearly all control lines are active low. The benefit of having the enable line is that we are able to disconnect a register without doing any physical uncoupling of links etc.
2 Read/write. The terms 'read' and 'write' are used to describe the direction of data movement. Loading data levels into a register is called 'writing' to it, and we 'read' the data to recover it.

### Using a register

The sequence is as follows:

1 The read/write line is taken to logic 0 to allow the register to receive data from an external source.
2 The enable control switches ON the tri-state buffers at the input to each flip-flop.
3 Data is written to each flip-flop and then the enable control puts the register to sleep until the next time it is needed.

### How long can the data be stored?

It will be stored until the power supplies are removed – either by an equipment fault or, more usually, by the system being switched off. The data does not deteriorate in storage.

## Shift registers

These are really ring counters without the final connection from the output back to the input.

They are called shift registers because the data is shifted from one flip-flop to the next each time the clock pulse occurs. The one we used for the ring counter shifted the data from left to right, and such a register would be called a 'shift-right' register.

The one in Figure 16.4 is a shift-left register. The last one in bit 7 drops off the end and is lost while, at the other end, a new bit is entered into bit 0 (Figure 16.4).

**Figure 16.4**
A shift-left register

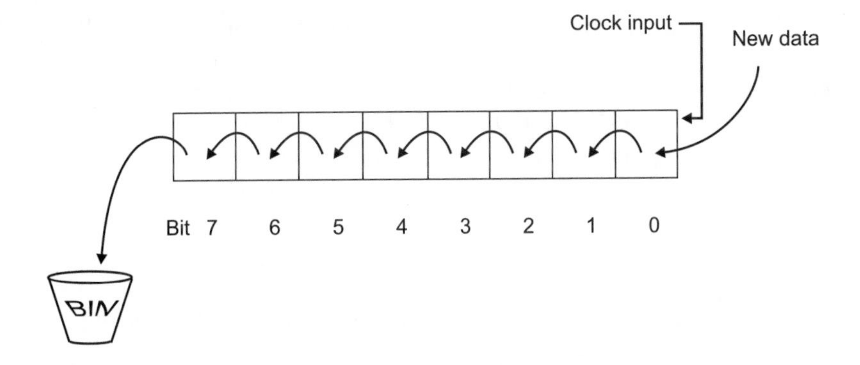


If we tie the input to a logic 0, each clock pulse will move the data one step and add a new zero at the right-hand end. The register will slowly have its data replaced by zeroes. We could load with ones if we wished.

## A real world use for a shift register

In Figure 16.5, it is controlling an automatic drinks dispenser. The customer inserts some money and presses any button of the eight available to obtain the drink required – but which button was pressed?

**Figure 16.5**
Using a shift register

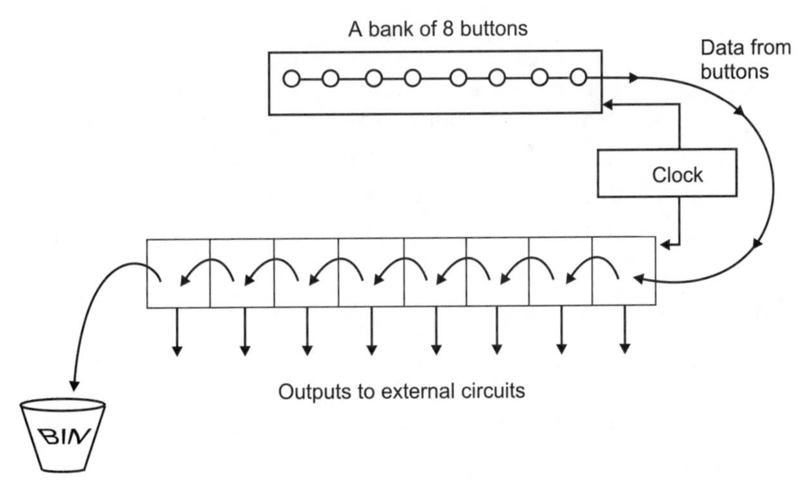


As a button is pressed the associated bistable data changes from logic 0 to logic 1, so to start with, we can assume no buttons are pressed and the response from each button is zero.

Along comes a customer who, having read the instructions, inserts some money, re-reads the instructions and stares at the buttons, eventually deciding to press one of them.

Pressing a button generates a burst of eight clock pulses, and the value of each button is loaded into the shift register. Once the button has been pressed the zeroes and ones corresponding to each of the buttons is loaded into the shift register. The output from each button is made available to external circuits, and one such circuit will be activated so that the chosen can drops down the chute.

## Rotate registers

These are modified versions of the shift registers and perform just like the ring counter.

As with shift registers, rotate registers can be made in rotate-right as well as rotate-left versions. In some cases, the same register can be used to rotate or shift in either direction.

The benefit of using a rotate rather than a shift register is that the data is not destroyed. We have seen that a shift register is progressively emptied as bits fall into the bin at the end. With a rotate register, the data is not changed. If we rotate left say, six times, we only have to rotate right six times to recover all the original data.

### Inputs and outputs

If, as in Figure 16.5, we load the data in serial form and send it out in parallel to another circuit, this type of register is referred to as a serial-in, parallel-out or SIPO register.

There is an obvious alternative by loading in a parallel way and then starting a clock pulse to move the data out in serial form. This provides a parallel-in and serial-out action, called a PISO register.

We can load in parallel and read in parallel in a PIPO, or use a serial method to load and read the data out, which is a SISO.

## 74XX194

This is a 4-bit universal register able to load in serial or parallel fashion and to shift in either direction. Outputs are available in parallel or serial format.

Let's look at the pins first in Figure 16.6.

### Outputs

The outputs of the shift register are shown as $Q_A$, $Q_B$, $Q_C$ and $Q_D$. If a parallel-out shift register is required we would use all of these outputs at the same time, but if we wanted a serial output we would use just one of them – probably $Q_D$. There is nothing to prevent us from using both the serial and the parallel outputs at the same time.

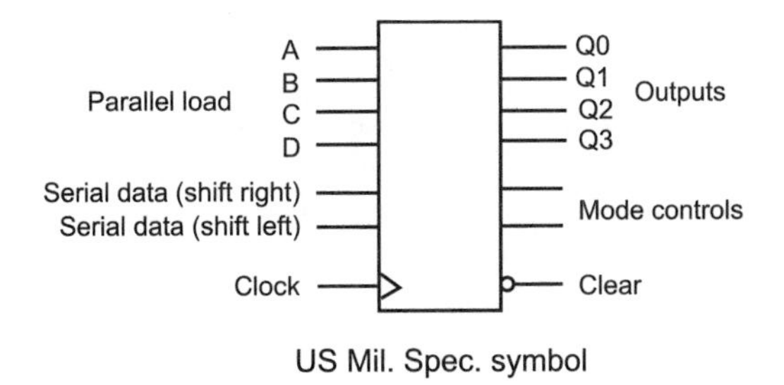


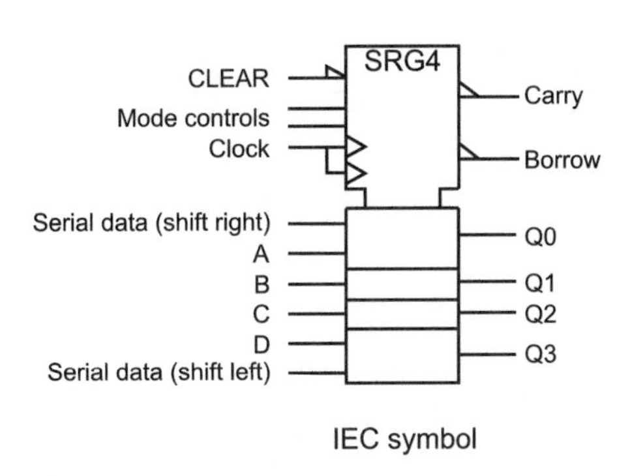


**Figure 16.6**
A universal shift register

### Inputs

The inputs follow much the same pattern. If we apply our data to all the inputs A, B, C and D we have a parallel-in register, but if we need a serial-in shift register we have a choice of applying the data to either the serial data shift-left pin or to the shift-right pin according to the direction of shift that we need.

### Mode controls

The input pins only provide the data to be shifted – they do not tell the register to shift in a particular direction. This is the job of the mode control pins. The register has four modes of operation controlled by the logic levels applied to the mode pins, as described in Table 16.3.

**Table 16.3** Mode control of the 74XX194A

| *S1* | *S0* | *Operation* |
|---|---|---|
| L | L | memory state |
| L | H | shift right |
| H | L | shift left |
| H | H | parallel load |

### The truth table

The complete and frightening truth table is shown in Table 16.4. We will have a look at it line by line.

**Table 16.4** Truth table of the 74XX194A

| *Inputs* | | | | | | | | | | *Outputs* | | | |
|---|---|---|---|---|---|---|---|---|---|---|---|---|---|
| | *Mode* | | | *serial* | | *parallel* | | | | | | | |
| *clear* | *S1* | *S0* | *Clock* | *L* | *R* | *A* | *B* | *C* | *D* | $Q_A$ | $Q_B$ | $Q_C$ | $Q_D$ |
| L | X | X | X | X | X | X | X | X | X | L | L | L | L |
| H | X | X | L | X | X | X | X | X | X | $Q_{A0}$ | $Q_{B0}$ | $Q_{C0}$ | $Q_{D0}$ |
| H | H | H | ↑ | X | X | a | b | c | d | a | b | c | d |
| H | L | H | ↑ | X | H | X | X | X | X | H | $Q_{AN}$ | $Q_{BN}$ | $Q_{CN}$ |
| H | L | H | ↑ | X | L | X | X | X | X | L | $Q_{AN}$ | $Q_{BN}$ | $Q_{CN}$ |
| H | H | L | ↑ | H | X | X | X | X | X | $Q_{BN}$ | $Q_{CN}$ | $Q_{DN}$ | H |
| H | H | L | ↑ | L | X | X | X | X | X | $Q_{BN}$ | $Q_{CN}$ | $Q_{DN}$ | L |
| H | L | L | X | X | X | X | X | X | X | $Q_{A0}$ | $Q_{B0}$ | $Q_{C0}$ | $Q_{D0}$ |

↑ = leading edge of clock pulse; a, b, c, d = level of data applied to parallel inputs; $Q_{A0}$ etc. = previous values of $Q_A$ etc. (memory state); $Q_{AN}$ etc. = previous level of $Q_A$ etc. before the most recent clock pulse.

**Note**: the high to low transition of S1 and S0 must only occur during the high state of the clock.

Line 1: The clear is taken low and, being active low, it resets all the outputs to a low state. Notice that the clock is marked as X, which is the 'don't care' state, because the clear can over-ride the clock and all the other inputs.

Line 2: With the clear high but the clock low there is no clock input, so no changes occur and the register is in its memory state. The outputs stay at their previous values. This is indicated by writing the zero after the Q states, e.g. $Q_{A0}$.

Line 3: With clear high and the two mode controls at S0 = S1 = high, it is in its parallel load mode. We can see that the parallel inputs lettered a, b, c and d are transferred to the $Q_A$, $Q_B$, $Q_C$ and $Q_D$ outputs at the moment of the $0 \rightarrow 1$ transition of the clock pulse. The register is not shifting the data along so the serial inputs are not used – hence the Xs in the table.

Lines 4 and 5: With S1 low and S0 high the resister is acting as a shift-right register. Each shift occurs on the leading edge of the clock pulse and the new data is read in from the serial data shift-right input pin.

This new data appears at the $Q_A$ pin, and all the previous data levels are shifted one place to the right. Like all shift registers, the original data that was in $Q_D$ has fallen off the end and has been lost. The previous levels are written as $Q_{AN}$ etc.

Lines 6 and 7: These follow the same pattern as lines 4 and 5 except that the direction of the shift is now towards the left. The data is loaded into the serial data shift-left pin and when the clock pulse occurs, the new data will enter at $Q_D$ and each of the previous data levels will move one place to the left until the original value of $Q_A$ is lost.

Line 8: With S1 and S0 both low, the shift register ceases to function and reverts to a memory state. When in the memory state, all the inputs are irrelevant and can be any value and are therefore shown as X on the table. The outputs are shown in the 'old' state and given the symbols $Q_{A0}$ etc.

### Unused inputs

When the shift register is being used as a serial load register the parallel inputs are not used; likewise, when parallel loading occurs, the serial inputs are irrelevant. All the unused inputs can be left disconnected and floating since their voltage levels do not matter. This is a very unusual state of affairs with digital circuits.

We return to registers in Chapter 17, but for the moment we will look at some new counting codes.

## Binary coded decimal (BCD)

If we want to represent a denary number in a binary form, we have more ways than the obvious choice of a straight conversion. There are a dozen or more ways of representing a denary number, each with their own advantages and disadvantages. If the input to a circuit gives rise to a binary code at the output, we have what we call a 'binary coded decimal' or BCD.

## Weighted codes

Many BCD codes are weighted. This means that each bit has a value depending on its position. Denary is weighted, since a number 5 in the right-hand column is five, but in the next column it means 50 and in the next 500 and so on.

There are other weighted codes, such as the 5421 code in which $5_{10} = 1000_{5421}$, the 2421 code in which $5_{10} = 0101_{2421}$, and the XS-3 (excess-3) which is the 8421 code plus 3, hence the excess-3. In this code, $5_{10} = 1000_{XS\text{-}3}$.

It is a good idea to make it clear which code is being used by adding suffixes, like we do with binary and denary.

## The Gray code

We met this in Chapter 9. It is a binary code, but it is not a weighted binary code. This means that each bit in the code does not have the same binary value.

You may remember that it starts: 0000, 0001, 0011, 0010. If this represents the count 0, 1, 2, 3 it cannot be weighted, for, whatever the weight of each bit, the value of 0011 must be greater than 0010; therefore it cannot be counting up.

The Gray code is often used for encoding data from computer-controlled machinery or to measure shaft angles or rotation speed. Its main advantage is that since only 1 bit changes from count to count, then if one change is missed, the maximum error can only be a single count. The Gray code shaft encoder disk is shown in Figure 16.7. This particular one divides the rotation up into angles of 22.5°. The disk is fixed to the shaft. The light and dark areas are detected by reflecting light from the surface, and the values are read from the centre of the disk by a series of light detectors.

**Figure 16.7**
A Gray code shaft encoder

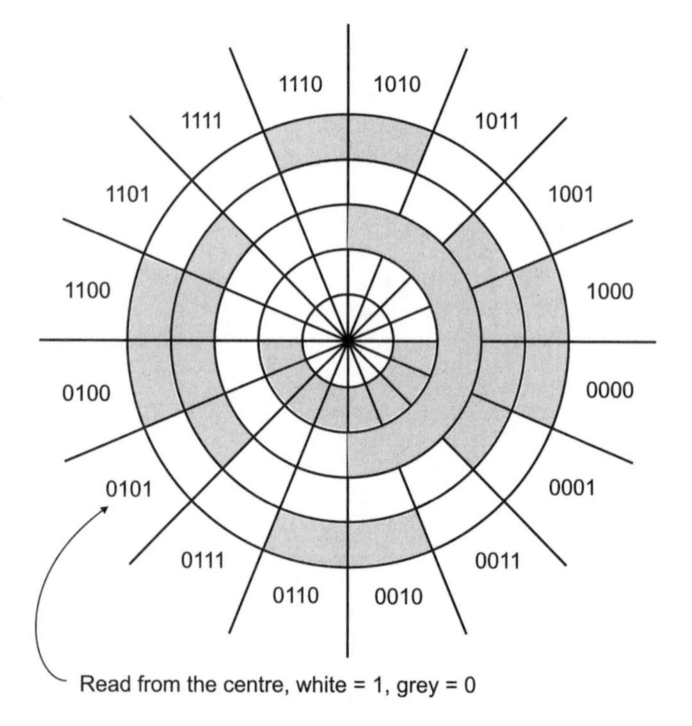


## The '8421' or 'standard' code

Conversion between binary and denary is not terribly easy, as we saw in Chapter 2.

One way of making it easier is to convert each denary number into a 4-bit binary number and simply string them together rather like we did when we were converting between binary and hex.

The numbers 0–9 are easy:

| Denary | Binary |
|---|---|
| 0 | 0000 |
| 1 | 0001 |
| 2 | 0010 |
| 3 | 0011 |
| 4 | 0100 |
| 5 | 0101 |
| 6 | 0110 |
| 7 | 0111 |
| 8 | 1000 |
| 9 | 1001 |

The next number is $10_{10}$, and this will not be written as the straightforward binary equivalent of $1010_2$. The denary digits 1 and 0 are treated separately. The 4-bit code for 1 is 0001 and the 4-bit code for 0 is 0000, so we can join them up to create the BCD equivalent of $10_{10}$ to be 0001 0000. In situations where it may otherwise cause confusion it can be written as $0001\ 0000_{8421}$ or, rather less correctly, as $0001\ 0000_{BCD}$. It makes it easier to read if the bits are separated out into blocks of four. This is all a bit like converting from binary to hex, isn't it?

The '8421' just reminds us that the values of each column are weighted as 8, 4, 2 and 1. In denary, the columns are weighted in powers of 10.

A BCD code is often used in circuits such as in electronic voltmeters, frequency counters, calculators, petrol pumps and other places where we want a denary display as an output.

## BCD counters

If we organize a decade counter to produce a 8421 BCD at its output, we can string a series of them together to produce as many digits as required – as shown in Figure 16.8.

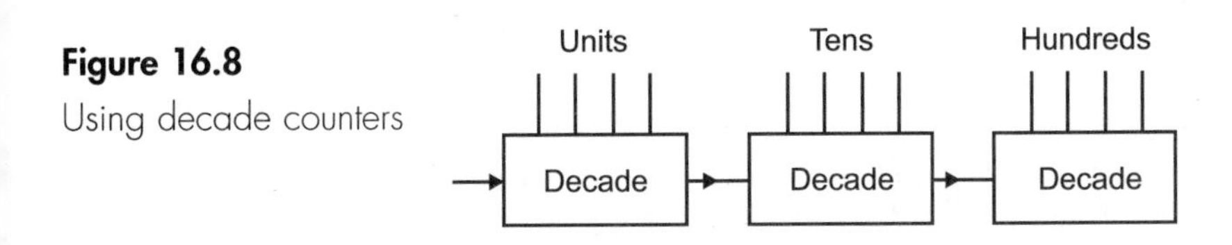


**Figure 16.8**
Using decade counters

We have looked at the 74XX390 bi-quinary version of a denary counter, but the same chip can be reconfigured to give us an 8421 BCD output. This is shown in Figure 16.9 (pin-out in Figure 15.10) and the count sequence is shown in Table 16.5.

**Figure 16.9**
The 74XX390 as a BCD counter

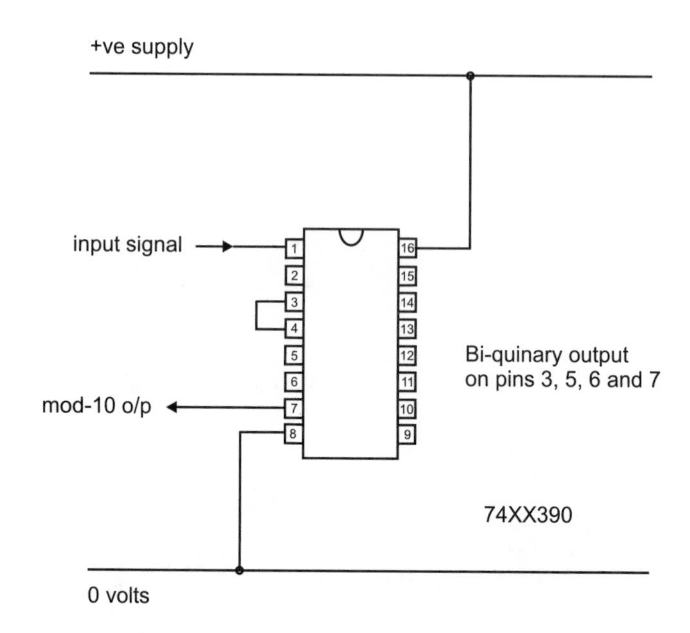


**Table 16.5** An 8421 BCD count from the 74XX390

| Count | Outputs | | | |
|---|---|---|---|---|
| | $Q_D$ | $Q_C$ | $Q_B$ | $Q_A$ |
| 0 | L | L | L | L |
| 1 | L | L | L | H |
| 2 | L | L | H | L |
| 3 | L | L | H | H |
| 4 | L | H | L | L |
| 5 | L | H | L | H |
| 6 | L | H | H | L |
| 7 | L | H | H | H |
| 8 | H | L | L | L |
| 9 | H | L | L | H |

The outputs are now taken in the expected order

## Quiz time 16

In each case, choose the best option.

**1 An example of a unweighted BCD code is the:**

(a) Gray code.
(b) BCD code.
(c) 8421 code.
(d) 5421 code.

**2 Four bistables are required to count up to eight using:**

(a) a ring counter.
(b) an 8421 BCD counter.
(c) a Johnson counter.
(d) an asynchronous counter.

**3 If a 4-bit register is loaded with the binary value 1111 and shifted three counts to the right and then three to the left, the data stored will now be:**

(a) 0011
(b) 0001
(c) 1111
(d) 1000

**4 Two bi-quinary counters can be interconnected to provide a count of:**

(a) 25
(b) 250
(c) 8
(d) 40

**5 If the active-low enable input to a register was taken high, all data stored in the register would:**

(a) be reset to zero.
(b) remain at their previous value.
(c) be switched to a high logic level.
(d) toggle.

# 17

# Digital devices

In Chapter 16 we looked at registers which were used to store binary information, usually in groups of 4 or 8 bits at a time. Once stored, we could play tricks like shifting or rotating the information.

A memory chip is very similar.

## Memory

The function of a memory is to store information – almost the same as we said for the register. Generally, a register stores small quantities of data for immediate use, whereas a memory is designed for bulk storage of data but that is all it can do – no shifts or rotates. Some of them are very useful in that they can remember the data, even when the power is turned off, for up to 7 years.

This ability to remember data after the power is switched off is the dividing line between the two main types of memory. If the memory loses its data when the power is switched off, then we call the memory RAM or volatile memory. If it can hold on to the data without power, we call it ROM or non-volatile memory (volatile means 'able to evaporate'). This is seen in Figure 17.1.

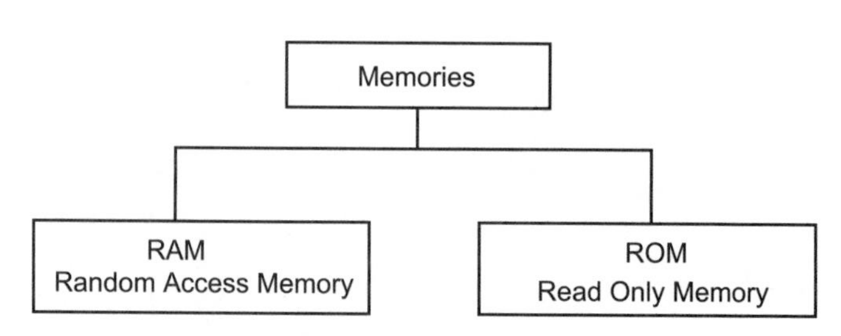


**Figure 17.1** Two types of memory

## RAM

Since RAM is so similar to our registers, we will make a start here.

Inside the chip, there are a large number of registers – hundreds, thousands or millions, depending on the size of the memory. Incidentally, when we are referring to memories, we use the word 'cell' instead of register even though they are exactly the same.

So, in each of the internal cells there is a 4-, 8-, 16-, 32-, or 64-bit register made from a line of bistables. The registers, or cells, are arranged in rows and columns so we can access the data easily. Figure 17.2 shows the register layout in a very small memory containing only 16 cells or locations, each of which can hold 4 bits and is given a memory number or address. By today's standards, this is an itsy-bitsy-too-small-to-mention register.

**Figure 17.2**
The layout of cells in a memory

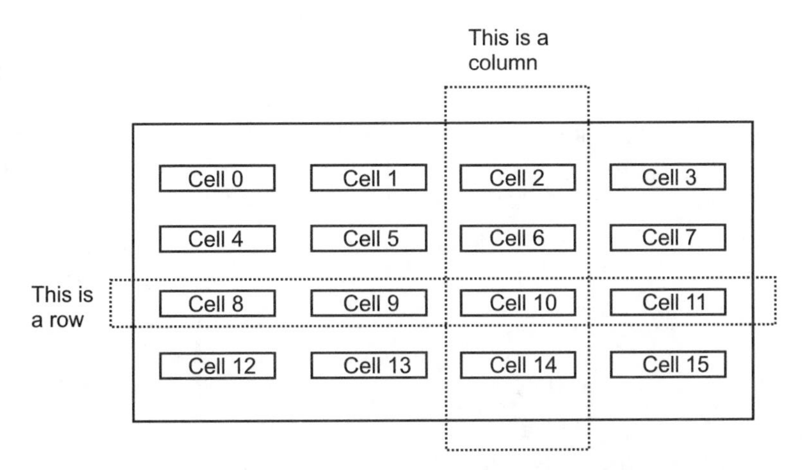


## Accessing memory

Loading data into a register is called 'writing' data to a memory, and we talk about data being 'read' from the memory.

Each location in a memory is given a number, called an address. In Figure 17.2, the 16 locations of memory would be numbered from 0–15 or, in binary, $0000–1111_2$.

The cells are formed into a rectangular layout, in this case a $4 \times 4$ square with four columns and four rows.

To use a cell, the row containing the cell must be selected and the column containing the cell must also be activated. The shaded cell in

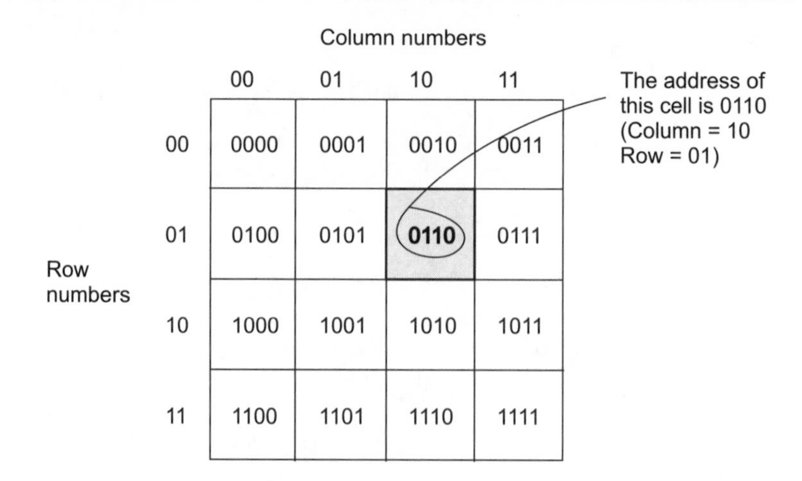


**Figure 17.3** Selecting a memory location

Figure 17.3 has the address 0110, which means that it is in row 01 and in column 10.

To access this cell we need to apply the binary address to the row and column decoders. When the address 0110 is applied, the first half of the address, 01, is applied to the row decoder, and the second half of the address is applied to the column decoder.

A decoder circuit is a small gate circuit which, when fed with the address of the location, is able to switch on the appropriate row and column.

The maximum number of locations that can be addressed will depend on the number of bits in the address. We have already seen that a 4-bit address can access 16 locations.

This is because $2^4 = 16$, so generally $2^n$ = number of locations if n is the number of bits in the address.

To take a more realistic example, if we had 20 address lines we would have $2^{20} = 1\,048\,576$ locations.

## Two types of RAM

RAM chips can be designed in two different forms called static RAM (SRAM) and dynamic RAM (DRAM), as seen in Figure 17.4.

## Static RAMs

These are the normal line of bistables. The problem with the flip-flop is that it draws current all the time because one of the two transistors in the totem-pole output stage is passing current all the time. It doesn't matter if we have just four bistables in a register, but an integrated

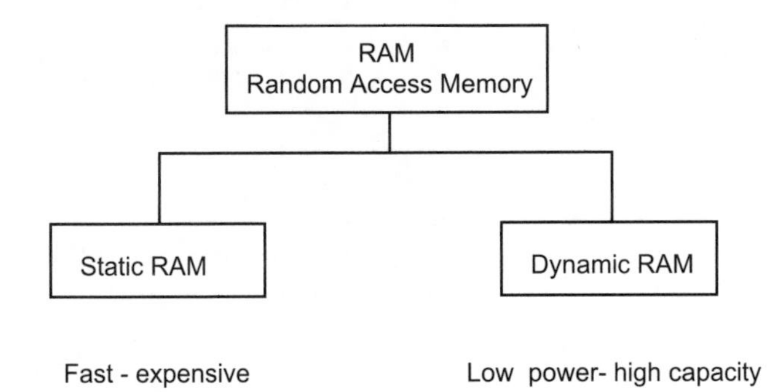


**Figure 17.4**
The two types of RAM

version with 100 000 bistables is another case altogether. The benefit of bistables is that they are very fast and are used where speed of access is important.

## Dynamic RAM

These store the information in capacitors, which are small components that store an electrical charge in the form of static electricity. They are called 'dynamic' owing to one of their drawbacks. In use, the electric charge stored in each capacitor leaks away because of the imperfect insulation. So after a little while the charge has to be replaced, otherwise the DRAM will be empty and all the stored information will be lost. This process of topping up the charge is called 'refreshing', and has to be performed at intervals of about 2 ms by a DRAM control circuit. To prevent any interference with the operation of the microprocessor system, the refreshing is done in the background whenever the DRAM is not being used.

Once the static charge is stored, no further current is required (except for refreshing), therefore less heat is being generated internally and we can pack more memory into a given space. We say it has a 'high packing density'.

### Memory organization

The size of a memory is always quoted as number of locations × bits stored in each, so this memory, which contained 1024 locations with 8 bits stored in each, would have an organization of 1024 × 8.

## Three types of ROM

All ROMS are used to store information on a more-or-less permanent basis. In use, the ROM can be read but new information cannot be stored in it. In other words, we cannot write to it (Figure 17.5). The choice of type depends on the number being made and how permanent we need the data to be.

**Figure 17.5**
Three types of ROM

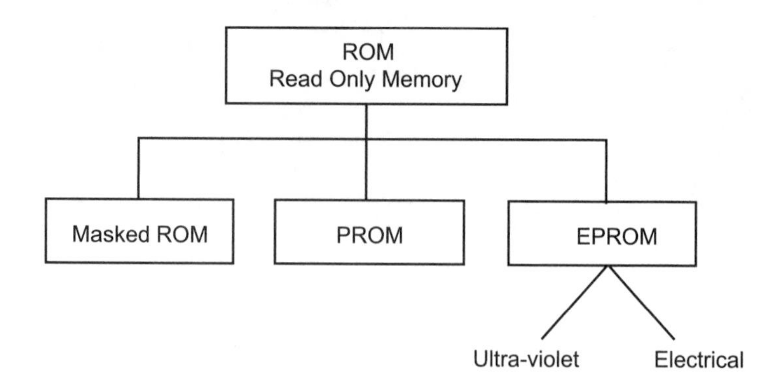


## Masked ROM

This one is cheap to manufacture if at least several thousand identical chips are required. It is only found in large-scale manufacture, such as in computers or test equipment. Once made, the data cannot be changed. We have to send our data to the ROM factory and the chips will be delivered with the data, right or wrong, permanently loaded.

## PROM (programmable ROM)

This chip is an attempt to get the price and convenience of a ROM but still be able to write our own data into it without committing ourselves to a large number of devices. It is ideal for situations where we may want to use between perhaps 100 and 1000 identical chips. Inside these chips there is a tiny fusible link for each data bit. The bit is set to binary 1 unless the fuse is blown, at which time it goes to binary 0. To do the blowing we use a 'PROM blower'. The data is loaded, the button is pressed and the equipment goes from location to location blowing the required fuses. If we make a mistake, we drop the chip in the bin and start again.

## EPROM (erasable programmable ROM)

As the name would suggest, this chip allows us to program it, then change our minds and try again. To erase the data there are two methods – ultraviolet (UV) light or electrical voltage pulses.

EPROMs are ideal for prototyping since it is so easy to change the data to make modifications.

The data is added by using a tool called an EPROM programmer. This is similar to the 'PROM blower'; indeed, it is often the same instrument.

The EPROMs arrive empty, which means all the locations hold a logic 1. The EPROM programmer is loaded with the required data and applies voltage pulses to load each location.

If we make a mistake, we must erase all the data and start again. We can have up to 700 attempts before the chip dies. It slowly gets more and more difficult to clear the old data, until we get tired of waiting and drop it in the bin.

The types of EPROM only differ in the method used for erasing the old data.

The UVEPROM is erased by shining a strong ultraviolet light through a small clear window in the top of the IC.

A specially constructed EPROM eraser provides the light. We pop the chip in, close the lid and switch on the timer. After a few minutes, the data is erased.

The eraser includes power cut-outs to ensure that we do not accidentally get exposed to the UV light. The UV light is not the fun stuff that makes our clothes glow at the disco; it is seriously nasty with a wavelength of around 254 nm. It can cause irreparable damage to our eyes in just a few seconds.

As an alternative, some EPROMs are erased and reprogrammed by electrical pulses. This type is called an EEPROM (electrically erasable programmable ROM).

### Accidental erasure

To ensure that the data is not accidentally erased during normal use, the programming voltage is always much greater than the normal use power supplies. Typically, EPROMs are made to operate from the usual logic supply of 5 V but they use a voltage of 12 V for programming. So, whatever we do with the pin connections, we cannot destroy the data. We could destroy the whole chip by reversing the power supplies, though!

Be careful with the programming voltages. We are nowhere near the point of having a standard programming voltage. A glance through the catalogue will show programming voltages varying from 12–25 V, all with a normal operating voltage of 5 V when reading data.

### Pin layout of an EPROM

Figure 17.6 shows the pin-out diagram for a 1 Mb (actually 1 048 576 bits) EPROM with an organization of 131 072 × 8 bits.

### Power supplies

The main power supplies to operate the chip are the +5 V applied to the V+ pin and 0 V on the GND (ground) pin.

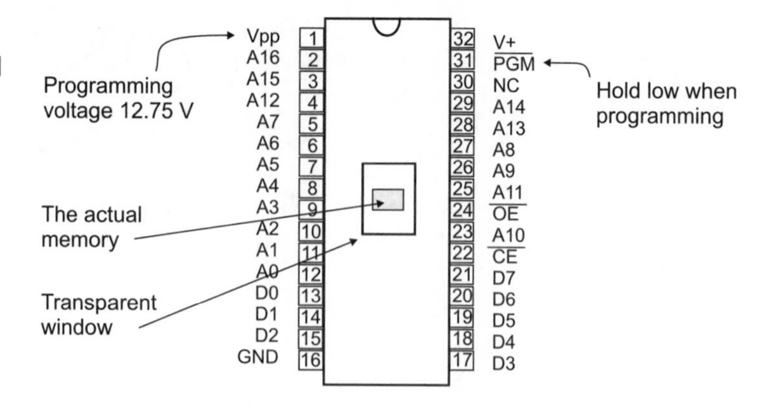


**Figure 17.6** The 27C1001 UVEPROM

To program the memory, the programming voltage is applied to Vpp. When not being programmed, it should be held at +5 V. The PGM pin should also be held low during this time.

## Address pins

Address pins are always numbered starting from A0.

The number of locations is given by $2^n$, so with 17 address lines (A0 to A16) the number of locations would be $2^{17} = 131\,072$.

## Data pins

Like the addresses, these pins always start counting from zero.

In the EPROM shown in Figure 17.6 they are abbreviated to D for data, and go from D0 to D7 – eight in all. Some manufacturers call them output pins and number them O0, O1, O2 etc. The output from these pins is the normal logic levels of near 0 or near +5 V.

## Control pins

1 The chip enable (CE), sometimes called chip select (CS), is the main on/off switch for the chip. It is usually active low, which means that the chip needs a logic 0 voltage to be applied to switch the chip on. This is indicated by a line over the CE. When the chip is switched off, it goes to sleep and the power drops with a reduction of about 150 times.
2 The output enable (OE) leaves the chip fired up but with its output disconnected from the data pins. How this is done, we will discuss in the next chapter. Disconnecting the output pins is very much faster than switching the chip off. Watch out for the line over the name to indicate the polarities required.

**Unconnected pins**

An unconnected pin is shown as NC (not connected) and is not used. It is physically separate from the internal chip and therefore has no effect on anything. It should be left unconnected.

## Converting binary to Gray code

Let's imagine we need to convert a series of binary numbers between $0000\,0000_2$ and $1111\,1111_2$ to the equivalent Gray code.

We have three options – two sensible and one silly. To dispose of the silly option first, we are not going to start from zero and count up in Gray code as we saw in Chapters 9 and 16 until we get to the number.

There are two better ideas. The first is to convert the codes by hand, and the second is to use an EPROM to help us.

## Converting by hand

See Figure 17.7 and follow the steps below. It seems strange to start with, but soon becomes quick and easy.

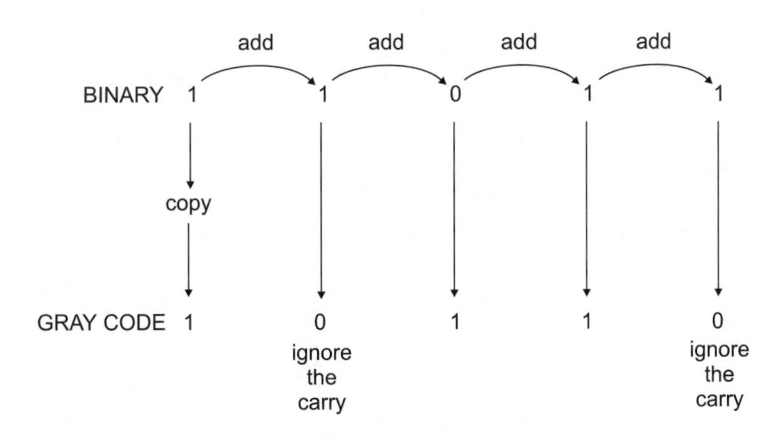


**Figure 17.7** Converting binary numbers to Gray code

**Method: Converting the binary number 11011 to Gray code**

Step 1: Copy the left-hand bit down to become the left-hand bit of the Gray code. In our example the MSB in the binary is 1, so the MSB of the Gray code is also 1.

Step 2: Add the MSB of the binary to the next bit. Ignore any carry that occurs and use the result as the second bit of the Gray code. In our example we have added $1 + 1 = 10_2$. Ignore the carry 1 and enter the 0 as the next Gray number.

Step 3: Add the binary bit in the second column to the one in the third column and copy the result into the Gray code. In our case, 1 + 0 = 1.

Step 4: The bits in the third and fourth columns give 0 + 1 = 1. Copy this answer down into the fourth column in the Gray code.

Step 5: The last two bits give $1 + 1 = 10_2$. As before, we ignore the carry and copy down the 0. This gives the final result of $11011_2 = 10110_{GRAY}$.

## Using an EPROM

EPROMs and other memory chips are often looked upon merely as a computer component to store programs and data, but they can be used individually to convert between codes of any type.

We could store a series of numbers, let's say all the binary numbers from $0000\ 0000_2$ up to $1111\ 1111_2$ and the corresponding 256 Gray codes, in an EPROM, one Gray code in each memory location starting at memory location 0000 0000.

To convert any binary number, say $10011101_2$, to Gray code, all we have to do is to use the binary number as an address by applying it to the address pins on the chip. This address will be accessed and the contents, which will be our previously stored answer, $11010011_{GRAY}$, will appear at the EPROM data output. We have to produce all the Gray codes in the first place but, having done this, we can recall any

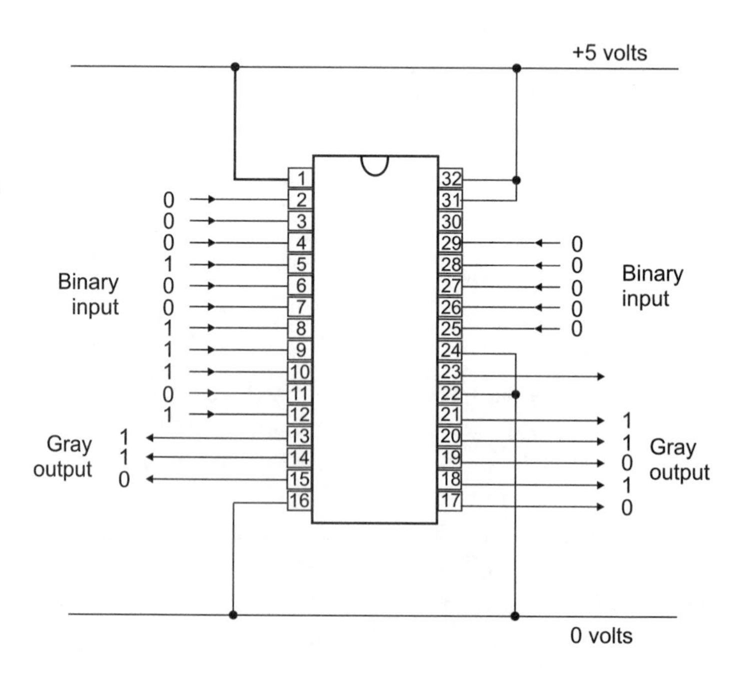


**Figure 17.8**

Binary is applied to the address pins and Gray code appears at the data pins conversion within about 150 ns, which is faster than working them out.

Clearly this system can be used to convert between any two codes. Just as a matter of interest, Figure 17.8 shows the above conversion being undertaken by an EPROM. Since we are converting an 8-bit binary number and there are 17 address pins, we just tie all the excess pins to 0 V to stop them floating and changing logic levels accidentally. Read the data sheets well before use, as the treatment of some pins differs according to the EPROM used.

## Converting Gray code into binary

Have a look at Figure 17.9 as we convert the Gray code $1011_{GRAY}$ into binary. The method is quite similar.

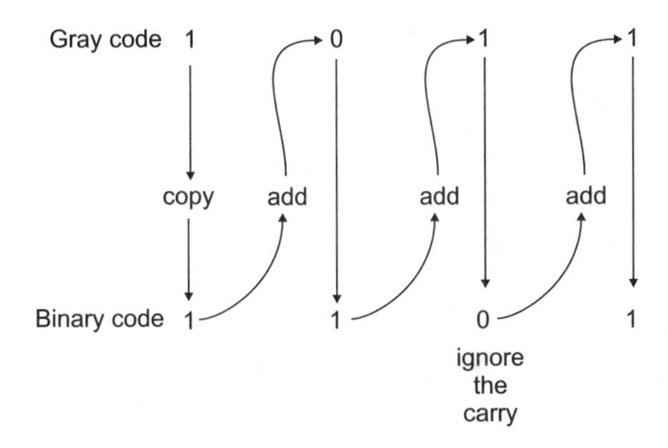


**Figure 17.9** Converting Gray code to binary

Step 1: Copy the first bit down to provide the first binary bit. This means that Gray code and binary always start with the same number.

Step 2: Add the first bit in the binary number to the second number in the Gray code. The answer is the second binary bit, 1 + 0 = 1.

Step 3: Add the second binary bit to the third Gray code digit and use this answer as the third binary bit. As before, just ignore any carries that occur. 1 + 1 = 10; use the 0 and ignore the 1.

Step 4: Add the third bit in the binary number to the fourth number in the Gray code. The answer is the fourth binary bit, 0 + 1 = 1.

This gives the final result of $1011_{GRAY} = 1101_2$.

## Adders

These circuits provide the addition needed to convert between Gray and binary codes as well as in other operations.

We can build the circuit ourselves or, more economically, we can use a ready-made chip.

## Building it ourselves

To design this logic circuit, or any other, we must start with a truth table or a Boolean expression as we saw in Chapter 7.

If we add two binary numbers, there are a limited number of outcomes. The answer may be 0 or 1, and there may be a carry-out.

The options are:

| A plus B | | Answer | Carry-out |
|---|---|---|---|
| 0 | 0 | 0 | 0 |
| 0 | 1 | 1 | 0 |
| 1 | 0 | 1 | 0 |
| 1 | 1 | 0 | 1 |

If we look at the Answer column, we may spot that we have an XOR gate and the carry-out is just an AND gate, so we can easily build the circuit shown in Figure 17.10. This circuit is called a half adder to distinguish it from the full adder, as we will see in a minute.

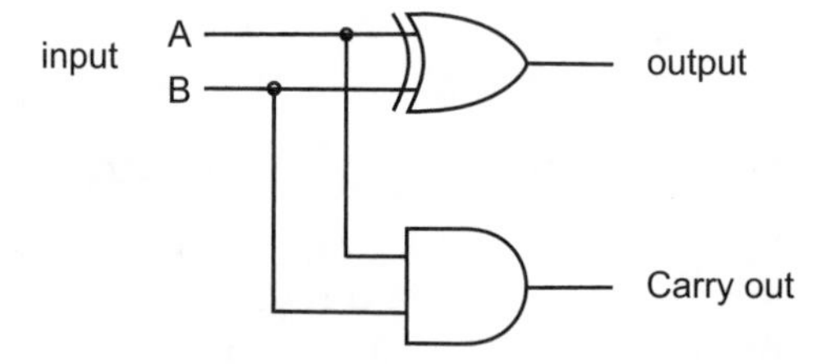


**Figure 17.10**
A half adder

The half adder is OK if all we want to do is to add 2 bits together, but life is never that simple. Binary numbers normally have at least 4 or 8 bits, so we need a way to join up (cascade) adders. When we try this, we must allow the possibility of the carry-out from one adder to be connected to the input of the next one.

A full adder takes care of this. The circuit has three inputs: A, B and carry-in.

**74XX283**

Apart from the pin-out, it is identical to the 74XX83a.

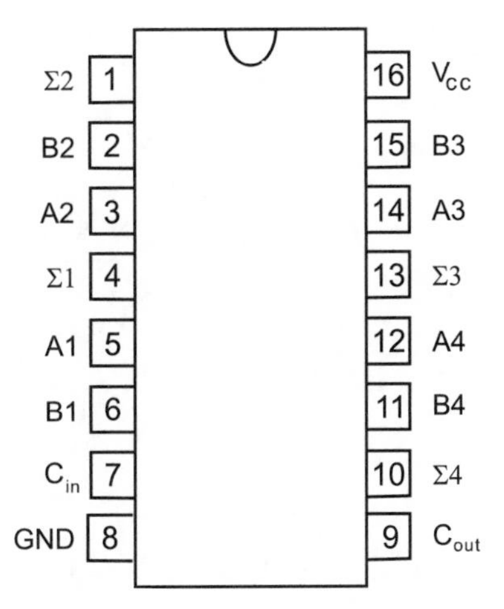


**Figure 17.11**
The 74XX283

It contains two 4-bit full adders referred to as A and B. The first bit of each number, A1 and B1, are added, together with the carry-in, and the answer is available at Σ1. This can be summarized as $A1 + B1 + C_{in} = \Sigma1$.

The next 3 bits are treated in exactly the same way. The final carry is available on pin 9 as $C_{out}$. Extra pins are expensive to add, and to avoid this only the first and the last carry are taken out to pins although inside the chip they are all connected to provide the correct arithmetic results.

## Displaying the results

### LED seven-segment displays

The display consists of a series of LEDs that can be activated separately to provide an alphanumeric display. These displays are normally sold in two flavours – common-anode and common-cathode – and in three colours.

By using a common pin we reduce the number of pins and the subsequent connections as in Figure 17.12, which shows a common-anode version.

A common-cathode version is exactly the same except that the LEDs are reversed. To illuminate a segment in a common-anode version, we simply take the voltage low so that the required current is flowing. This current needs to be limited so that about 2.2 V are dropped across the diode with the desired current flowing. Typically, each segment may draw about 20 mA and the current limiters may be about 150 Ω.

**Figure 17.12**
The common-anode version

Figure 17.13 shows a 74HC4511 chip driving a single display. This driver is able to run either common-anode or common-cathode displays. There are many different drivers such as the popular 7447A, which is an open collector device with active low outputs suitable for driving a common-anode device.

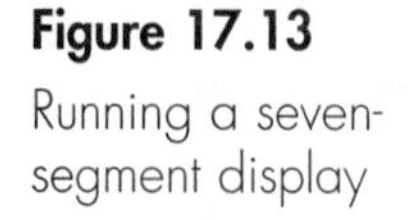

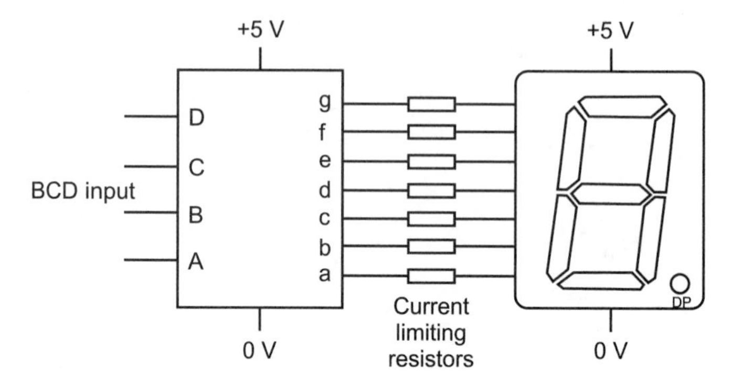


**Figure 17.13**
Running a seven-segment display

In general, common-anode displays are more popular as many logic circuits can sink more current than they can source.

### LCD

Liquid crystal displays, which are now common in calculators and other displays, offer some advantages over the LED displays. The main one is the extremely low operating current. Whereas each segment of an LED display takes about 20 mA, we can power four complete digits in an LCD from a 5 μA source at 3 V. They are easier to read in strong light conditions, and any shape or symbol can be displayed.

The driving is necessarily more complex, since a LCD requires an alternating square-wave input signal (it dies if fed with DC), but the 74XX4543 LCD driver will take care of this. With many characters appearing on the display, the input signals are necessarily more complex and are often microprocessor driven.

## Programmable logic arrays (PLA)

There is a whole collection of chips that are very similar. They go under a variety of names, which are also very similar. They are programmable logic arrays (PLA), field programmable logic arrays (FPLA), programmable logic devices (PLD) and programmable array logic (PAL).

Imagine buying several thousand AND gates, OR gates and some NOT gates, then soldering them onto a large board. Add plugs and sockets at all the inputs and all the outputs.

Whatever logic circuit we want to build, we can now do so by simply interconnecting whichever gates we want. Incidentally, the total cost of this lorry-load of gates is about the same as six AND gates. In logic, as in so many things, it certainly pays to buy in bulk.

All these gates are shrunk onto a single chip, and the required interconnections can be made by fusible links as in a PROM or by an electrical pulse as with EEPROMs. If we want large quantities they can be connected permanently during manufacture, like a ROM chip.

To distinguish between connections that are possible and those that are actually made, we use the symbols shown in Figure 17.14.

**Figure 17.14**

Connections that are made or are just possible

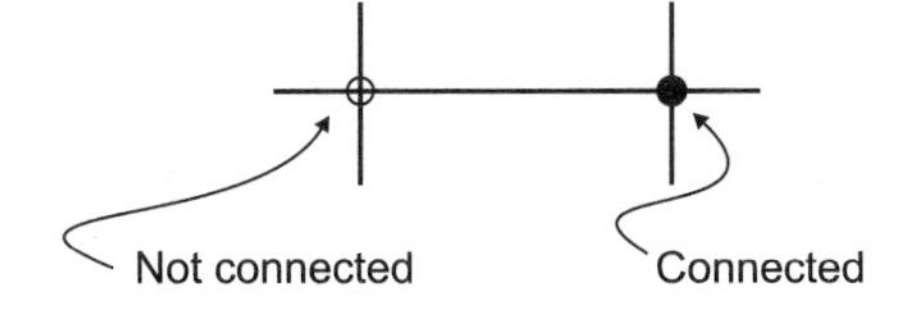


A very small sample of the full circuit is shown in Figure 17.15, which is enough to show the possibilities.

What gate have we built in Figure 17.15?

The output of the AND gate 'D' is $A\overline{B}$ and the output of gate E is $\overline{A}B$. These two outputs are applied to the NOR gate and provide the output $G = \overline{A\overline{B} + \overline{A}B}$.

The final output is an XNOR function.

**Figure 17.15**
What gate is this?

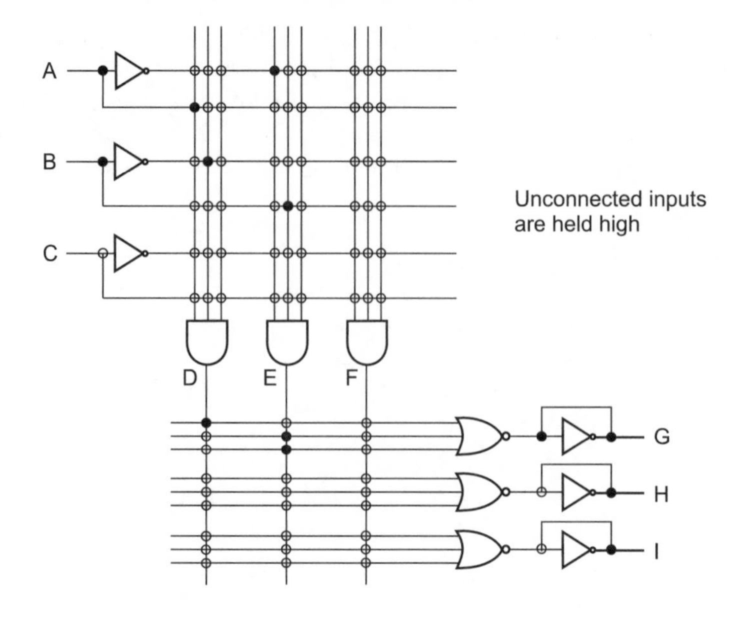


## A real PLA

Our example was deliberately simple, but was enough to show the possibilities. In a real chip there would be about 14 data inputs A, B, C, D . . . to replace our three shown. There are typically 96 14-input AND gates and eight outputs. This provides a mind-boggling number of possible interconnections.

Needless to say, computer-aided design plays a part in the organization of the connections. When designed, the connections are made by a piece of equipment similar to an EPROM programmer.

## Analogue-to-digital conversion (ADC)

We live largely in an analogue world. Some of our instruments are analogue and some are digital. The circuitry we use is partly analogue and partly digital, so it is not surprising that much conversion takes place between these two systems.

In essence, the analogue-to-digital converter (ADC) samples the analogue voltage from time to time and converts the measured value into a binary number. The speed and accuracy depend on the circuit used for the conversion. It is possible to build an ADC from discrete components, but no-one does this. It's just not economic.

There are several methods used in converters – here are four of them.

## Parallel or 'flash' converters

This is the fastest method of ADC, but can be expensive. Being very fast, a parallel converter also samples the incoming analogue signal for a very brief moment and is therefore largely unaffected by a changing input voltages. The sampling speed is governed by a clock input, and can be over 20 000 000 samples per second (20 MS/s).

A series of reference voltages are set up inside the IC and the input voltage is applied to each comparator at the same moment, as in Figure 17.16.

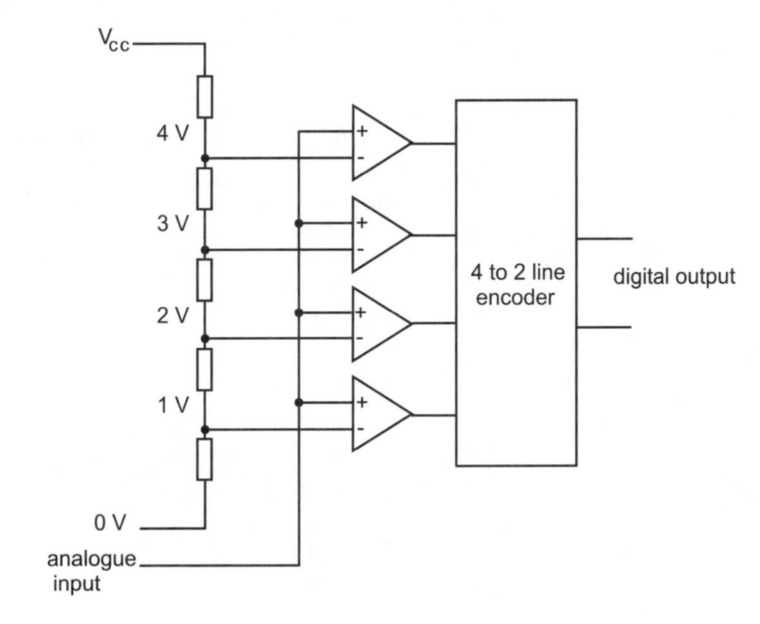


**Figure 17.16** A parallel or flash ADC

The comparators have two inputs. One is a reference voltage, which is compared with the input voltage applied to the other input. If the input voltage is less than the reference voltage, the output is logic 0. If it is greater, it will produce a logic 1.

Let's assume that we apply a voltage of 3.5 V. The first three comparators with reference voltages of 1 V, 2 V and 3 V will produce a logic 1, and the highest comparator will still provide a logic 0 at its output. The four-line to two-line decoder will represent the number of 'high' outputs as a binary number on its output lines. A more typical number of comparators would be between 16 and 1024 to give more steps in the digital output for a given analogue input, thus increasing the accuracy.

In our example the comparator reference voltages were spaced at 1 V intervals, so anywhere between just over 3 V and just under 4 V will give the same digital output. Having more reference levels can reduce this error, called quantization error.

The levels are converted into a binary number with between 4 and 10 bits to accommodate the 16–1024 input levels. A typical IC for this purpose is the eight-line to three-line encoder 74XX148.

## Single ramp

This is the opposite end of the spectrum: cheap and slow.

The input voltage is compared continuously with a single reference voltage inside the chip. The internal reference voltage increases at a known rate until it exceeds the incoming voltage. The time taken obviously depends on the magnitude of the voltage to be converted. To measure this time, an internal clock is started coincident with the start of the ramp voltage and is stopped by the switching of the comparator in Figure 17.17. It converts between 5 samples/s and 1000 samples/s, which is awful compared with the flash converter.

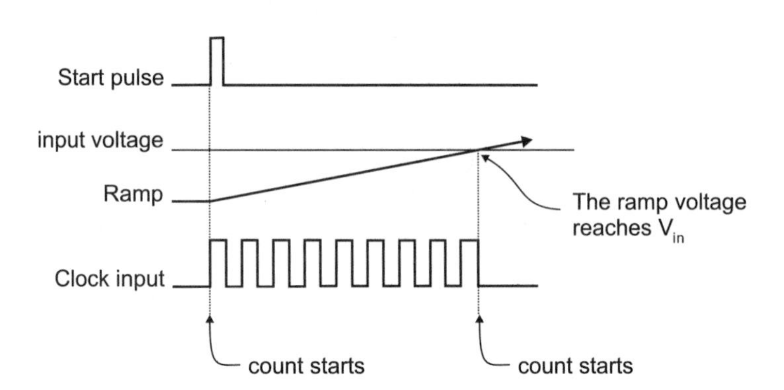


**Figure 17.17** Single slope ADC

## Dual ramp

This type is often found in instruments such as digital voltmeters.

It overcomes two main problems of the single slope ADC, caused by the inaccuracy of the clock frequency that determines the final count and the non-linearity of the ramp voltage.

The unknown voltage is applied for a set number of counts to a ramp generating circuit in which the upward slope of the ramp is determined by the voltage applied. At the end of the count, the ramp circuit decreases the voltage at a known rate by using an internal reference voltage and the same counter. When the ramp voltage gets to zero, the count stops.

$$\text{The input voltage was } \frac{\text{number of up counts}}{\text{number of down counts}} \times \text{reference voltage.}$$

### How about an example?

Assume that the fixed count used is nine, as in Figure 17.18 (it will be between 1000 and 260 000 in real life).

**Figure 17.18**
Dual slope ADC

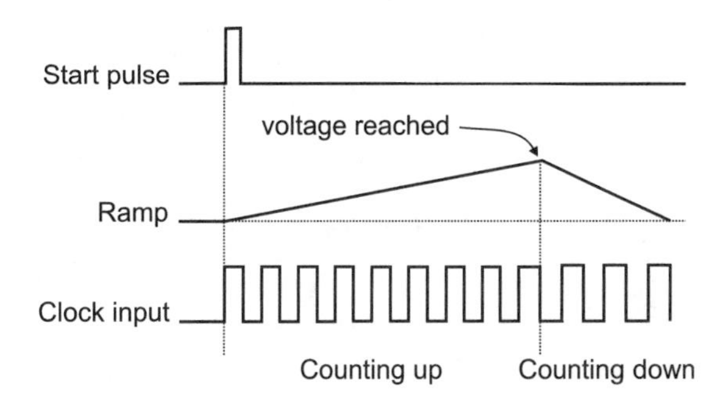


We apply the unknown voltage, and during the count time the ramp reaches a particular value. At the end of the count we reset the counter to zero and use an internal reference voltage to ramp down to zero. Let's assume the internal reference voltage was 5 V and, in our example, the ramp reached 0 V after three counts.

$$\text{The input voltage was therefore } \frac{9}{3} \times 5 = 1.66\,\text{V}.$$

## Successive approximations

This is a popular design which, although slower than the flash converter, is considerably faster than the dual slope circuits.

It generates a range of reference voltages, each of which is half the value of the next in line. It then adds these reference voltages to 'home in' on the required digital output value.

To demonstrate the principle of operation, we will use an over-simplified 4-bit version. The idea is shown in Figure 17.19.

If the input signal applied was 5.8 V, the sequence could be as below.

Step 1: Switch A is closed and the input voltage of 5.8 V is compared with the reference voltage of 4 V. The input voltage is larger so the

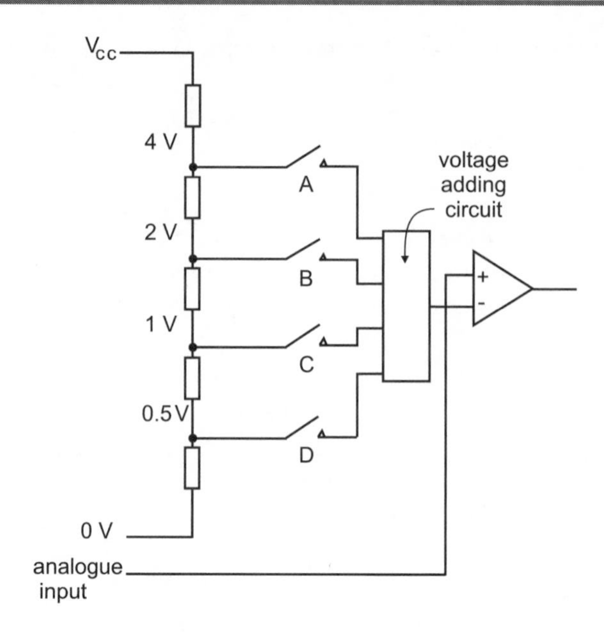


**Figure 17.19** Successive approximation ADC

comparator provides an output of binary 1, which is the most significant bit of the output.

Step 2: Switch B is closed and the combined voltage of 4 V + 2 V is applied to the comparator. This total of 6 V is rejected by the comparator since it is greater than the 5.8 V input signal. The comparator provides a binary 0 as the output value of the next bit.

Step 3: Switch C is closed and the reference input to the comparator now tries 4 V + 1 V = 5 V. This is less than 5.8 V so the output is another binary 1.

Step 4: The last switch is closed, creating a total of 4 + 1 + 0.5 = 5.5 V. The input is greater than this, and this last 0.5 V is accepted and a final bit = 1 is added to the digital output.

The final digital output is therefore $1011_2$.

If we return to a real-life size of 8–12 bit outputs, we only need 8–12 clock pulses occupying from 10–50 μs. Slower than the parallel converter, but much faster than the slope varieties.

## Digital-to-analogue conversion (DAC)

Conversion in this direction is rather easier. In essence, each bit in the binary number is allocated a voltage and the final outcome is achieved by simply adding the voltages together.

For example, if we had a 4-bit binary number the weightings would increase in steps of two as we moved from right to left across the columns. If we chose to make the LSB equal to 0.25 V, then the other columns would have the values 0.5 V, 1 V and 2 V.

A binary count of 1010 would provide an analogue output of 2 + 0 + 0.5 + 0 = 2.5 V.

An operational amplifier generates these voltages, with the input resistance value doubling as we go from bit to bit. Have a look at Figure 17.20.

The function of an operational amplifier connected in this way is given by the formula:

$$\text{output voltage} = \text{input voltage} \times \frac{-\text{R1}}{\text{R2}}.$$

When the first switch is closed,

$$\text{output voltage} = \frac{-10\,\text{k}\Omega}{50\,\text{k}\Omega} = 2\ \text{V}.$$

The next switch would result in 1 V at the output then 0.5 and 0.25 V for the remaining two switches.

The switches are controlled electronically by the binary value applied to the circuit. When more than one switch is closed at the same time, the voltages are added at the output.

A binary input of $1111_2$ would close all the switches, giving a maximum output of 2 + 1 + 0.5 + 0.25 = 3.75 V.

The design problem with this circuit lies with the resistors. As the number of inputs increases, the range of resistor values increases

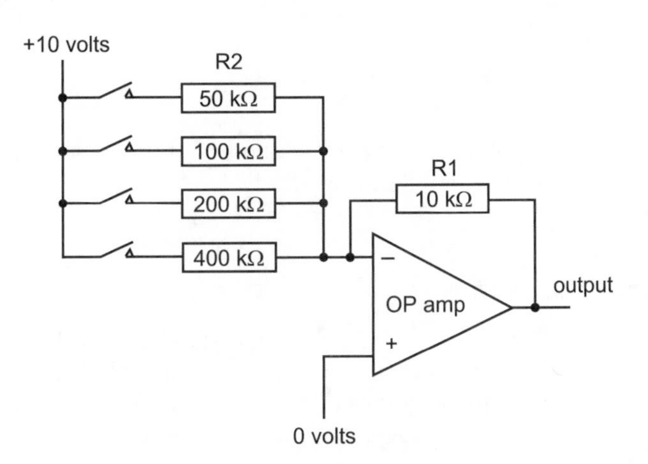


**Figure 17.20** A digital-to-analogue converter

rapidly. With our 4-bit converter in Figure 17.20, the ratio of the highest resistor to the lowest is 8:1.

Ready-made converters are usually 8 or 12 bits, with resistor ratios up to 4096:1. This is a very wide range bearing in mind that all resistors must have the same degree of accuracy.

Commercially available DACs normally use an improved resistor system called an R-2R pattern or an R-2R ladder. The advantage of this method is that, although more resistors are used, there are only two values, one being twice the value of the other. So we could use just 10 kΩ and 20 kΩ, for example. This means that it is very easy to maintain the required accuracy of the resistors.

Figure 17.21 shows a 4-bit DAC using an R-2R ladder to provide a direct contrast with the previous type in Figure 17.20. The switches perform exactly the same function as previously, i.e. putting the right-hand switch over to the binary 1 position (to +10 V) will result in –2 V appearing at the output. If we want +2 V we can always invert it with another op amp or a transistor.

**Figure 17.21** Easier to build, harder to calculate

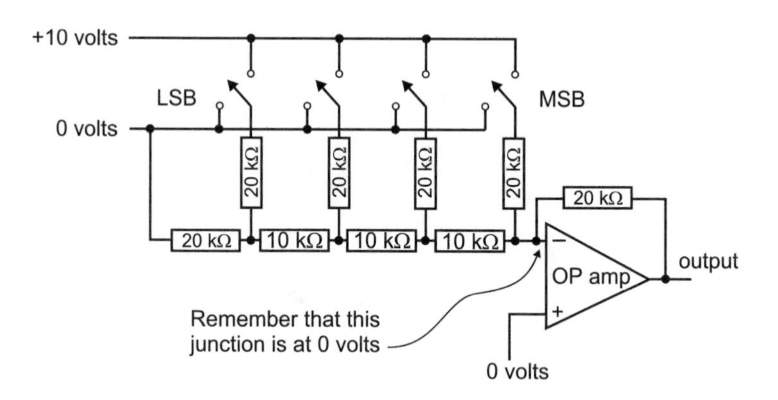


It would provide a few minutes of fun to confirm the calculations – or just accept that it works!

## Microprocessors

A microprocessor is an integrated circuit that accepts binary inputs, usually between 4 bits and 64 bits at a time. These inputs are either instruction codes telling it to perform a function like ADD, or they are data to be used or sent to the output connections. On its own it is pretty useless, so it is surrounded by RAM and ROM chips to hold the data to be used and store the results.

## How does it know what to do?

We have to provide it with a list of instructions called a program. How to actually follow these instructions is stored inside itself as a set of internal instructions. This internal code is the most commercially sensitive part of the design.

To add two numbers, the sequence of events will be something like this. We tell it to load the first number, usually from a RAM chip or an EPROM. It loads the second number. Finally it adds them and then stores the answer in a RAM chip for later use.

Everything must be split into tiny steps as the microprocessor has a very limited number of instructions, but it can carry them out at relatively high speeds.

## What is there inside the microprocessor?

Inside, the most modern and largest microprocessors have up to 9 000 000 transistors, generally organized into registers to shift data around.

They are classified by the number of bits that can be entered and sent to the output at the same time. This varies from the 4-bit microprocessors that run our dishwashers, microwave ovens and video recorders through 8, 16 and 32 bits to the 64-bit monsters that live in our computers.

Remember that, to do anything useful, the microprocessor must be surrounded by other devices like RAMs, ROMs, EPROMs DAC, ADCs and power amplifiers, power supplies and all sorts. It's very much like comparing an engine with a car. An engine sitting in the workshop is a long way off driving to work.

**Note**: Microprocessors are explained in the companion book *Introduction to Microprocessors,* by the same author.

## Programmable interface controllers (PICs)

These are self-contained systems that have a microprocessor together with the necessary surrounding chips to provide an output that is able to control external devices. It is a (sort-of) computer, but is geared towards controlling machinery, external devices and processes in an industrial setting. It must be programmed in much the same way as a computer.

## Quiz time 17

In each case, choose the best option.

**1 The first bit of a Gray code is always:**

(a) 1.
(b) the opposite value to the first bit of the binary equivalent.
(c) the same as the first bit of the binary equivalent.
(d) 0.

**2 When the power is switched off, data is lost from:**

(a) a RAM.
(b) an EPROM.
(c) a ROM.
(d) An EEPROM.

**3 Dual ramp is a type of:**

(a) R-2R.
(b) DAC.
(c) UVEPROM.
(d) ADC.

**4 The fastest form of analogue to digital converter is called a:**

(a) single ramp.
(b) parallel converter.
(c) binary up-counter.
(d) register.

**5 A full adder includes:**

(a) an adder, but it has no carry facility.
(b) a carry-out and a carry-in.
(c) a carry-out but no carry-in.
(d) a carry-in but no carry-out.

# 18

# Transmission of digital data

## Tri-state outputs and tri-state buffers

We have seen that the output of a gate can be of a totem-pole design, with the two output transistors in series acting like a couple of switches.

Some digital circuits, particularly those using microprocessors and large scale digital circuits, have outputs with three alternative states. The output can be logic 1, logic 0 or disconnected, as in Figure 18.1. In the tri-state condition, the chip is, in effect, isolated from its output.

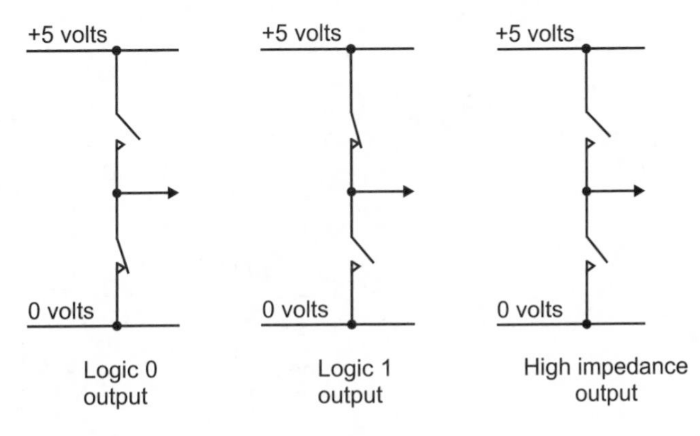


**Figure 18.1** The three states of a tri-state

'TRI-STATE' is actually a trade name owned by the National Semiconductor Corporation. Over time it has become a generic term, in the form of 'tri-state', for what should strictly be called 'three-state' devices.

To disconnect the chip there is a pin or pins called output enable ($\overline{OE}$), or just enable ($\overline{E}$). When these pins are held at a low logic state the chip behaves normally, providing a logic level at all the outputs that depend on the input conditions. If we take these pins high, the chip outputs are disconnected. It does this by disabling a series of tri-state buffers immediately before the data pins.

There may be several enabling pins on a chip, and not all will be active low though this is the most common.

## Tri-state buffers

A buffer is a circuit that has been added to provide isolation between two circuits, a bit like a firebreak. It reduces the chance of unwanted noise and oscillations escaping from the output back to the input and causing distortion or other mischief.

A buffer often provides voltage or current amplification, and sometimes it inverts the signal. Buffers are also available in the tri-state form.

If we want to switch off a data stream, it is undesirable to use a normal on–off switch. The timing of interruption is impossible to control accurately and the switch may bounce. It is far better to use a tri-state buffer, which will connect or disconnect at the moment when the voltage is applied to the enable input.

Buffers are available in integrated form as quad, hex or octal buffers; that is, four, six or eight of them in a single chip. Figure 18.2 uses the 74XX125 quad buffer.

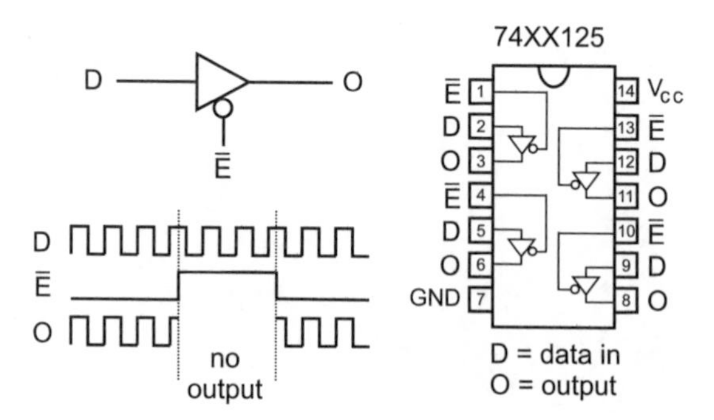


**Figure 18.2**
Switching with a tri-state buffer

## Bus working

Bus working is a way of reducing the number of interconnecting wires by using tri-state devices. It is very similar to the telephone system in that all our telephones use the same 'trunk' route, which acts just like a bus. The rather impractical alternative would be to have a separate wire leaving our house to each person in the world that we may wish to contact.

We are using a bus in Figure 18.3 in which the seven-segment display can be controlled by either of the two driver circuits.

**Figure 18.3**
On the buses

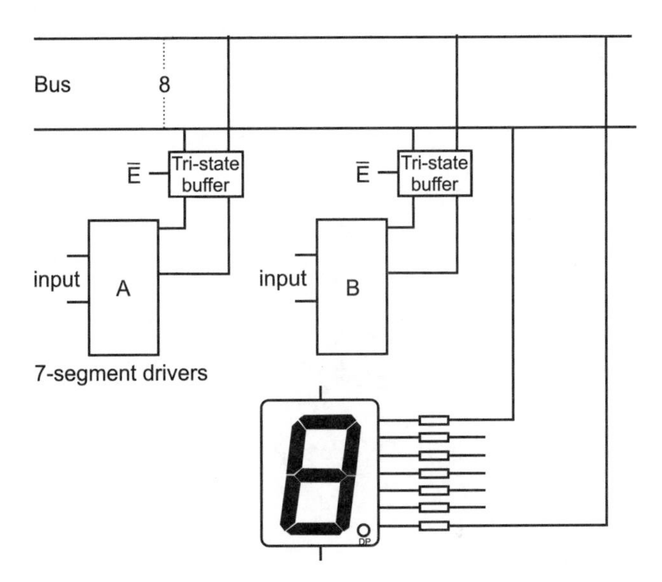


If we wish to use input B to provide the signal for the display, we must isolate driver A by switching off the tri-state buffer with a high logic level on its enable input. Taking its enable input to a low state connects the buffers for input B. In this way, either of the two sets of data can use the same bus connections. For simplicity the figure shows only two circuits using the bus, but in reality there can be as many as we like providing that only one input is switched on at the same time.

On the diagram, the bus consists of only eight connections. They can all be drawn on the diagram, or we can just show the number of connections included by adding a label as in the figure. By the nature of the circuitry that is using this bus it will be a one-way route, but in other cases data can be carried in either direction as required.

A long bus often employs bus drivers that are usually tri-state buffers designed to provide high current levels. A typical example is the 74XX125.

If a bus is too long it begins to display odd effects. What these are and how long is too long, we will see in the next section.

## Transmission line speed and impedance

A transmission line is any connector, like a wire, cable or copper track on a printed circuit board, that is used to convey energy.

Any connection includes a small amount of resistance and inductance and some capacitance, as in Figure 18.4. The main point here is that if electrical energy is going to travel along the wire, then it follows that the energy must be contained within the transmission line at some time. Therefore the characteristics of the transmission line will have an effect on the signal.

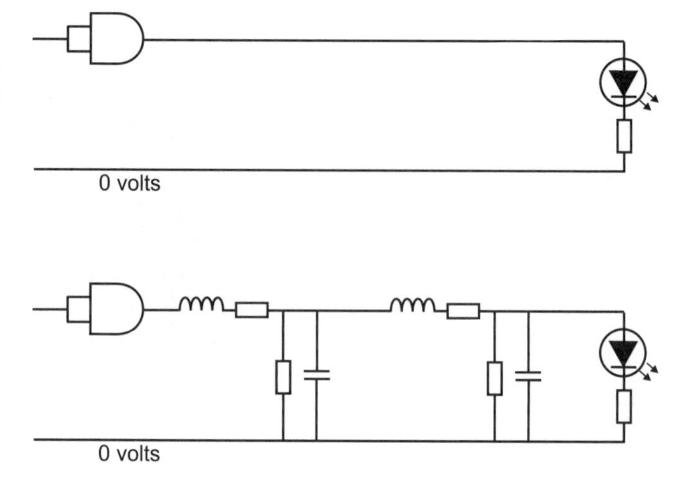


**Figure 18.4**
A transmission line is more than just a conductor

One effect is that these characteristics determine the speed at which the energy can move.

The formula is:

$$\text{velocity} = \frac{1}{\sqrt{\text{LC}}} \text{ metres per second}$$

where L is the inductance per metre and C is the capacitance per metre. For a typical transmission line it works out to be about 17 cm/ns (light travels at $3 \times 10^8$ m/s or 30 cm/ns in free space).

Another effect is that the inductance and capacitance determines the impedance of the line. Impedance is the ratio of voltage/current, just like in Ohm's law. If we ignore the effects of resistance, the impedance of any length of line, called the 'characteristic impedance' is given by the formula:

$$Z_0 = \sqrt{\frac{L}{C}} \text{ ohms}$$

where L and C have the same values as previously.

### So what's the point of all this?

If the energy is travelling along a transmission line which is terminated by an impedance equal to the characteristic impedance, the pulse remains square and the rise time is unaffected (Figure 18.5).

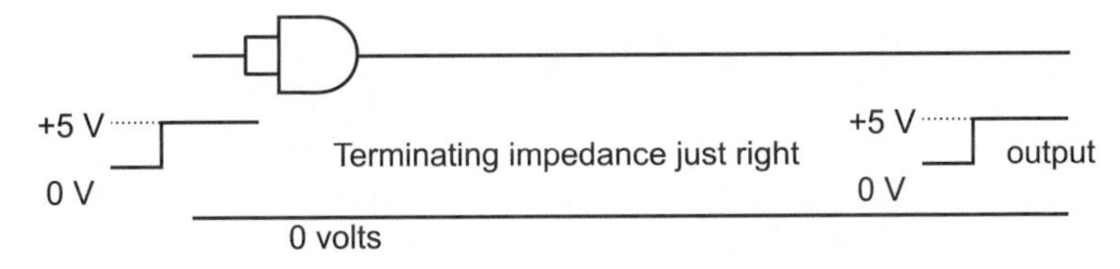


**Figure 18.5**

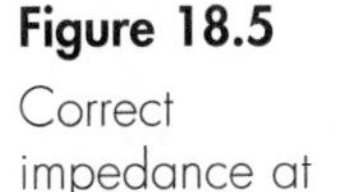
Correct impedance at the far end

However, if it hits an impedance at the far end which is different to the line impedance, some of the energy is reflected from the end of the line. This is just like waves bouncing off the end of a bath. The returning energy may then be reflected from the other end and bounce to and fro along the line, being progressively absorbed by the resistance as it goes.

If the end impedance is too high, the voltage will bounce back in-phase and will add. The effect is shown in Figure 18.6. Notice how 'ringing' occurs with the voltage overshoot and an undershoot possibly low enough to change logic levels. It loses energy on each reflection, and will stabilize at +5 V after a few reflections.

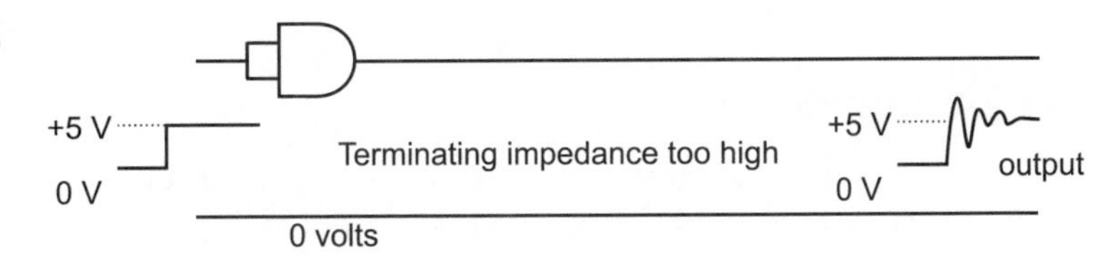


**Figure 18.6** Reflecting off a high impedance

If the terminating impedance is too low, it will have the effect of slowing down the increase and thus increasing the rise time. This is shown in Figure 18.7.

**Figure 18.7**
End impedance too low

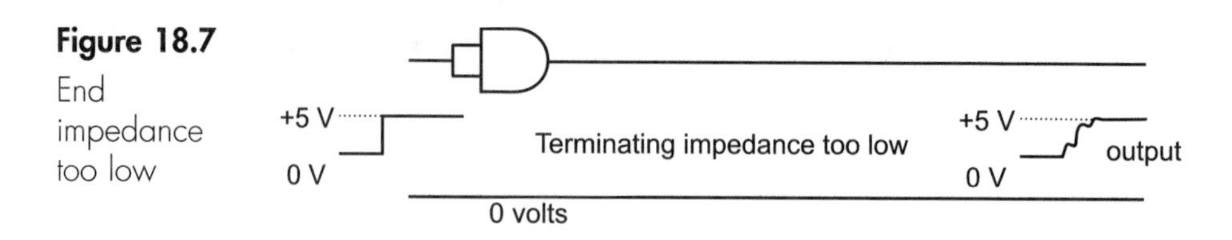


### The good news

If a connector is short enough, we don't have to worry about all these reflections. But how short is short enough?

We will have problems if the reflected energy gets back to the driving gate after the change of voltage level is completed. The longer the delay, the worse the problem. So we don't want fast rise times and long transmission lines.

Let's look at an example. The 74LS series has typical rise times of 5 ns and fall times of around 2.5 ns. Taking the faster of the two, the reflection must get to the end and back in less than 2.5 ns. This means it could cause problems if it took more than 1.25 ns to go each way along the line.

If we use our typical figure of 17 cm/ns, the line can cause problems if it is over $1.25 \times 17 = 21.25$ cm – and the longer the more likely the problems.

As a rule of thumb, if a line is over 300 mm in length and rather odd things keep happening, it may be a transmission line effect.

### A solution

Since the amount of reflected energy depends on the difference between the characteristic impedance of the transmission line and the terminating impedance, one first line of defence is to terminate the line with an impedance that roughly matches most lines – around 100 Ω.

When ringing occurs the undershoot causes most problems, and to counteract this effect the 74F1016 can be used. It is designed to remove the undershoot by using Schottky diodes. This chip has 16 terminations, and only requires inputs and a 0 V connection (no $V_{CC}$).

## Noisy environments

A conductor, such as a transmission line, will suffer from crosstalk if it is running parallel to another conductor carrying data. It will also suffer from electrical noise, whether it is natural as in a thunderstorm or artificial as with motors, fluorescent lighting, machinery etc.

Using a screened cable will reduce but not entirely eliminate the noise. However, there is an easy way to totally eliminate all of these problems, and that is by using an optic fibre.

The military have another problem. One effect of a nuclear explosion is an amazingly intense electromagnetic pulse that will easily kill off normal digital networks and circuits. For most of us, the fact that our hi-fis and mobile phones won't work after a nuclear attack has a worry value of about 0.000001 on a scale of 1–10. To the military mind, however, it is still important to have enough control over our communications to be able to organize the elimination of the other country.

Again the answer is optic fibres.

## An optic fibre

An optic fibre is a very thin, ultra-clear length of glass. We have known for over a hundred years that communication by light is possible, but we neither had the light source nor the glass for long distance communications. In 1960 we developed the ruby laser, which gave us the light source, but we still didn't have the glass to shine it through. We poured millions of pounds, dollars and yen into research, and after many years silica glass was developed. The clarity of this glass is nothing short of amazing. If the ocean was made of it we could see right down to the bottom of the deepest part. We could make a pane of window glass several kilometres thick, and it would be as clear as a sheet of normal window glass.

An optic fibre is made of two layers of this glass; the inner one, called the core, is used to carry the light, and the outer one, the cladding, is there to prevent the light from escaping. The combined thickness of the layers is about 125 μm or approximately the same thickness as two pages of this book.

For very short ranges of a few metres we can use a clear plastic fibre such as is used in some hi-fi units. This is cheaper and easier to use than glass, but it has considerably higher losses.

To build a digital communication system we need a light source, like a laser for long distances or an LED for short ranges, a length of optic fibre, and a light detector to convert the flashes of light back into voltage pulses. This is done by a PIN diode, which is really just a fast-

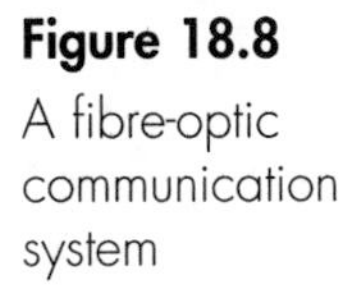

**Figure 18.8**

A fibre-optic communication system

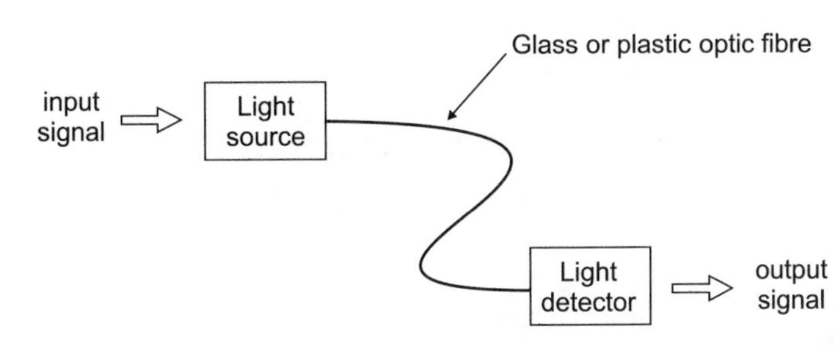

acting photocell that looks like an LED but works the opposite way round – we put light in and get voltages out.

A communication system is shown in Figure 18.8.

The story continues in the companion book *Introduction to Fibre Optics*, by the same author.

## Opto-isolators or opto-coupler

These integrated circuits are something like a very short optic fibre system. Inside the IC is a light source in close proximity to a light-sensitive switch of some kind. There are many different types on the market, which differ in speed, voltage and internal design, but the 6N136 shown in Figure 18.9 is designed to be compatible with the 74LS and CMOS 4000 families.

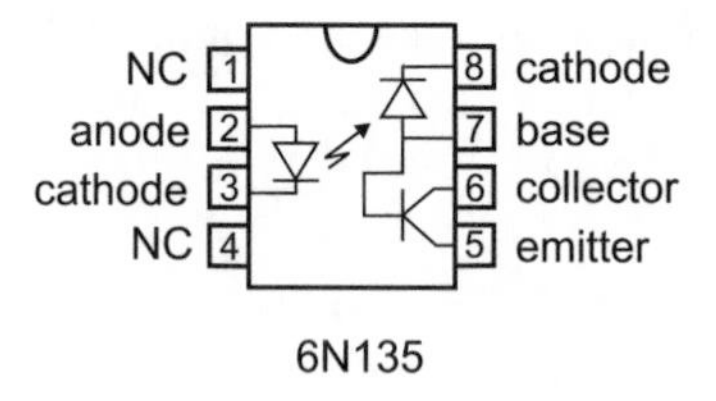


**Figure 18.9**
An opto-isolator

In this one, the input circuit provides a current that passes through the light source on pins 2 and 3. The resultant light causes the output diode to conduct, and this switches on the transistor that provides the input to the next circuit.

The advantage of using opto-isolators (also called opto-couplers) is that there is no DC voltage connection between the input and output circuits, so they can be used to link circuits where the voltages are incompatible. This one offers electrical isolation up to 2500 V.

### Quiz time 18

In each case, choose the best option.

**1 Optic fibre can be made:**

(a) only of silica glass.
(b) of copper.
(c) of sea water.
(d) of glass or plastic.

**2 Ringing occurs when the terminating impedance of a transmission line is:**

(a) equal to the characteristic impedance of the line.
(b) greater than the characteristic impedance of the line.
(c) equal to 100 Ω.
(d) less than the characteristic impedance of the line.

**3 Tri-state devices are:**

(a) essential for bus working.
(b) opto-isolators.
(c) not used in a RAM chip because they are too slow.
(d) able to provide output logic levels of 0, ½ and 1.

**4 Which of these situations is MOST likely to cause a problem with a transmission line?**

(a) A line length of 200 mm.
(b) A slow switching speed and a long transmission line.
(c) A high switching speed and a long transmission line.
(d) An impedance of less than 100 Ω.

**5 An opto-isolator:**

(a) requires an input of 2500 V.
(b) is an essential part of a fibre optic system.
(c) can be used to connect circuits in which the voltages are different.
(d) is normally used to terminate a transmission line.

# 19

# Data on the move

One of the simplest examples of data transmission is a device designed to inform us that we have a visitor. It is activated by a simple spring-loaded switch near the door. At the transmitting end a switch is closed and an electrical signal is passed along a wire to some receiving equipment, which may be a bell, chimes, music generator or a light. The receiving apparatus may well be reinforced by the sound of a barking dog crashing its way towards the door!

In this case the data being sent is simply a rectangular voltage pulse, but it is still a form of data because it is conveying information to us. This is not much different to the data being sent from the laser system in a CD player to the amplifier and then to the speakers in a hi-fi system.

## The ASCII code

Another short-range, slow-speed communication system used in a computer or other system to send alphabetic data is the ASCII code. ASCII stands for American Standard Code for Information Interchange. By short-range, we mean just that – a few metres, and then not with any degree of certainty.

It uses a 7-bit code that allows $2^7$ or 128 different characters to be transmitted. Some of these are for housekeeping jobs, like the tab key, the return key or the ESC (escape) key, and others provide the normal letters, numbers and symbols that appear on all keyboards.

### 7 bits or 8 bits?

The ASCII code is actually specified as a 7-bit code with 128 characters plus an extra bit designed as a parity bit, which makes it an 8-bit code. We will look at what parity is all about in a minute or two.

## An ASCII transmission

Assume we send the word 'It'. We start by looking up the ASCII equivalent for each of the letters: I = 49H, which in 8421 BCD format binary is $0100\ 1001_2$; then t = 74H or $0111\ 0100_2$.

So the data to send is 0100100101110100 – but how do we send it? The simplest way would be just to connect a piece of wire from the transmitter to the receiver. However, the reliability of this system would be poor. If, for instance, a moment's electrical noise removed a single bit, all the following ones would be in the wrong place and our word could be totally scrambled. How does the receiving device know whether one of the bits is wrong? This can be achieved by a parity check.

## Parity

If an announcement simply said '12.34' and nothing else, most people would either ignore it, stare at the loudspeaker or miss it entirely.

As a form of communication, this is not very efficient.

If it said 'the next train will depart at 12.34', everyone would understand the message.

The first attempt was very efficient in terms of the number of words used, but is unlikely to be an efficient communication because many people will not receive the message.

In the second attempt, we have used seven words to make sure that the three important ones get through. This is called adding 'redundancy'. The more redundancy we add, the more certain the message but the slower and less efficient the communication system.

Data being returned from space probes use very high levels of redundancy, over 96%, which allows for correction of really scrambled signals due to the extremely low power levels involved.

We can use parity for alerting us to the possibility of an error in a stream of data, or in some cases we can both detect and correct the error.

In its simplest form, we take a group of bits in a transmission – 4 or 8 bits are normally used, though the idea is applicable to all other values. In this example we will look at the first 4 bits in our letter 'I', which is the 4-bit group 0100.

At the transmitting end, we add an extra bit on the end – either a 0 or a 1 – to make the total number of 1s an even number. In this case there is only a single 1 in the group, so we add an extra 1 on the end. The data now reads 01001.

At the receiving end, if the data has been mutilated and now reads 01101 a quick count will show that there is now an odd number of 1s and so an error has occurred.

This simple approach can be easily fooled. If there are two errors there may still be an even number of 1s, which will be passed as correct. Another disappointment; if it shows an error, we cannot tell which bit is wrong and therefore cannot correct it. We can send a request back to the transmitter saying 'send that last group again', but this assumes that the transmitter and the receiver are in communication with each other.

To send our message 'It' we need to send a total of 16 bits, and with a little cunning we can not only spot an error but can automatically correct it.

Just follow these steps to send the binary data 0100 1001 0111 0100.

Step 1: Rewrite the data in the form of a square, using 4 bits at a time.

```
0   1   0   0
1   0   0   1
0   1   1   1
0   1   0   0
```

Step 2: Add parity bits.

Across the top row we have the numbers 0100, which includes a single 1. In this system, which we call 'even' parity, we add another 1 if necessary to ensure that there is an even number of 1s across the first row. As we have only a single 1, we add another 1 on the end.

It now looks like this:

```
0   1   0   0   1
1   0   0   1
0   1   1   1
0   1   0   0
```

The top row now has an even number of 1s. The next row has two 1s already, so we add a 0 to keep the total number of 1s an even number:

| | | | | |
|---|---|---|---|---|
| 0 | 1 | 0 | 0 | 1 |
| 1 | 0 | 0 | 1 | 0 |
| 0 | 1 | 1 | 1 | |
| 0 | 1 | 0 | 0 | |

The third row, with three 1s, will also be completed with a 1, and finally the last row will need another 1 to be entered.

The result is now:

| | | | | |
|---|---|---|---|---|
| 0 | 1 | 0 | 0 | 1 |
| 1 | 0 | 0 | 1 | 0 |
| 0 | 1 | 1 | 1 | 1 |
| 0 | 1 | 0 | 0 | 1 |

Step 3: Add parity downwards.

We now have five columns down the page, and we can add extra 1s in the same way to make the total number of 1s in each column an even number.

The fourth column already has an even number of 1s, so this one takes a 0. The other columns must have a 1 added. Notice that we also played the parity trick on the last column, which consisted of all the previous parity results.

The result is now:

| | | | | |
|---|---|---|---|---|
| 0 | 1 | 0 | 0 | 1 |
| 1 | 0 | 0 | 1 | 0 |
| 0 | 1 | 1 | 1 | 1 |
| 0 | 1 | 0 | 0 | 1 |
| 1 | 1 | 1 | 0 | 1 |

We have now got a total of 25 bits to be transmitted. This represents 16 bits of data and 9 bits added to check the accuracy of the data.

The final serial transmission is 0100110010011110100111101.

This means that 9 out of 25 bits (36% of the transmission) are not actual data and represent redundancy.

Let's see how it works. We will assume an error has occurred and one of the bits is received incorrectly, so here is the received transmission:

0100110010011110110111101

Step 1: Layout the data as a 5 × 5 square.

| | | | | |
|---|---|---|---|---|
| 0 | 1 | 0 | 0 | 1 |
| 1 | 0 | 0 | 1 | 0 |
| 0 | 1 | 1 | 1 | 1 |
| 0 | 1 | 1 | 0 | 1 |
| 1 | 1 | 1 | 0 | 1 |

Step 2: Check the parity in each row across the square.

Remember that we decided to use even parity, so each row and column should have an even number of 1s.

The first row has two 1s; this is even – OK.
The second row also has two 1s; this is even – OK
The third row has four 1s; this is even – OK.
The fourth row has three 1s; this is odd – an error has occurred.
The last row has four 1s; this is even – OK.

We now know that one of the bits in the fourth row has been received incorrectly – but so far, we don't know which one.

Step 3: Do the same for the columns.

The first column has two 1s; this is even – OK.
The second column has four 1s; this is even – OK.
The third column has three 1s; this is odd – an error exists in column three.
The fourth column has two 1s; this is even – OK.
The last column has four 1s; this is even – OK.

Step 4: Isolate the error and change the data.

The error occurs in the third column and the fourth row. Since we only have a choice of 0 or 1, we can confidently change the 1 to a 0 and recover the correct data stream:

| | | | | |
|---|---|---|---|---|
| 0 | 1 | 0 | 0 | 1 |
| 1 | 0 | 0 | 1 | 0 |
| 0 | 1 | 1 | 1 | 1 |
| 0 | 1 | **0** | 0 | 1 |
| 1 | 1 | 1 | 0 | 1 |

### A few notes

In this example we chose to use even parity; that is, we made each row and column have an even number of 1s. It would work equally well if we used odd parity by making the number of 1s an odd number.

It would also work just as happily if we counted the 0s instead of the 1s.

If more than one error occurs, it will warn us of an error but it will be unable to make any corrections. If you try it, you will see that it indicates four possible positions for the two errors and nine for three errors.

## RS-232

So far we have simply connected two circuits with a piece of wire and shovelled some data bits down it, hoping that it might come out the other end without loss or mutilation.

We may be more demanding, wishing to send data as far as 15 m and want reliable communication between two pieces of equipment.

## What is RS-232?

RS-232 is a laid-down standard that specifies the use of a 25-core cable so that all pieces of equipment that use this standard can be simply plugged together and work with no delay or hassle. The plugs and sockets are standard, and the voltages and signals on each conductor are itemized. Nothing can possibly go wrong – until you try it, that is.

**Note**: RS-232 refers to a method of sending information. It can be any information that we wish. The ASCII code is one type of information that can be sent along an RS-232 link, but we could equally well transmit any other data by RS-232.

There have been several versions of RS-232, and the current popular version is RS-232C – so, strictly speaking, all our references to RS-232 should really read RS-232C. But despite this, almost everyone forgets to put the C on the end.

## The RS-232 signal

There are two things about the signal that jump out at us: the voltages are big and the polarities are 'wrong'.

Logic high is between –5 V and –15 V at the transmitter output. The voltage is usually about –12 V.

Logic low is between +5 V and +15 V at the transmitter output.

The receiving end must accept a slightly wider range of between 3 V and 25 V.

These wide voltage differences give a good immunity to noise.

If we were using an RS-232 method to transmit our word 'It', the data would be sent as in Figure 19.1.

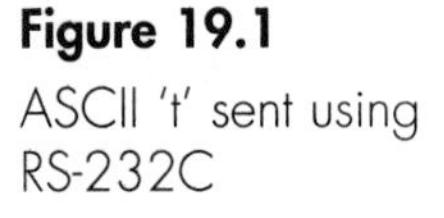
**Figure 19.1**
ASCII 't' sent using RS-232C

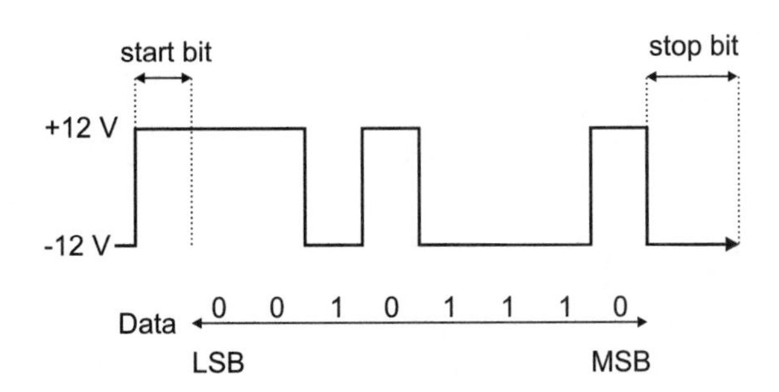

### How is it transmitted?

At the transmitter end of the communication link, the data is converted into RS-232 format and 1 bit of data is sent at each clock pulse.

Immediately before the first bit of data, a 'start bit' is sent. This is always positive-going, and warns the receiver that data will follow immediately.

After that come the data in reverse order – the least significant bit first.

Finally, there is a 'stop bit' to signify the end of the group of data.

Every bit of data is the same width except the stop bit, which is at least 1 bit wide, but there is no upper limit so the system can lie dormant waiting for the next data to be sent.

### Why are the start and stop bits needed?

For the data to be read at the receiving end, the data amplitude must be checked roughly in the centre of each data bit. This means that the receiver clock controlling the reading of the data must be going at much the same speed as the transmitter clock generating it. If this were not the case a bit might be read twice or missed altogether.

To keep the two clocks synchronized, the receiver clock is corrected every time the received signal changes from 0 to 1 or from 1 to 0; that is, on every edge of the signal. In many cases, as in Figure 19.1, we have several bits of the same value and no edges occur, so the receiver clock must be within a few percent of the transmitter rate to stay roughly synchronized. If we send a burst of data where the values are all 1s or all 0s, there will be no edges. To overcome this problem we use a start and stop bit of opposite signs, so that each character must generate at least two edges to keep the receiver clock on track.

### How fast is it sent?

The speed of transmission is measured in 'Baud' (Bd), which is the rate of the clock used to transmit the data.

In many systems like this one a single bit is sent by every clock pulse, which gives a transmission rate in Baud numerically equal to the number of bits per second.

The standard rates are 300 Baud, 600, 1200, 2400 and so on. The longer the line, the lower the maximum speed. RS-232 will run at up to 38 400 Bd, or even higher in good conditions.

## Generating the RS-232 signal

The signal conversion is always left to integrated circuits. In days gone by there were the two favourites numbered 1488 (TTL to RS-323) and

1489 (RS-323 to TTL), but these were slightly inconvenient because we had to provide +12 V and –12 V supplies.

Nowadays we have new chips like the MAX220, which is made by Maxim. The advantage here is that they only need a single +5 V supply and have an internal circuit to generate RS-232 voltages. Even so, they only draw a low current of about 2 mA.

A typical system is shown in Figure 19.2.

**Figure 19.2**
An RS-232 system

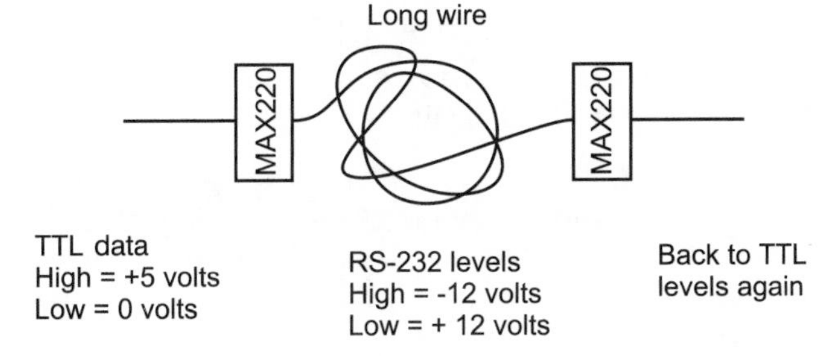


**So what are the problems?**

Very few pieces of equipment actually use all the specified 25 connections, and those that are used may be used for different purposes.

Equipment is divided into stuff called data terminal equipment (DTE) and other stuff called data communication equipment (DCE). The wiring for each type is different; some use a particular wire as an input and some use the same wire as an output. In most cases, we don't know what a piece of equipment actually is and the manufacturer doesn't tell us. It requires experimentation.

Plug them together. They may work. If they don't, more reading may be worth while – see *The Art of Electronics* (Horowitz and Hill, 1989).

## RS-485 and RS-482

These other two standards were designed for faster speeds and for use over greater distances – we are talking about 1 km and 10 MBd. They also allowed more than one transmitter to use the same transmission line and more receivers to be connected. In the case of RS-422 we were looking at 10 receivers and 1 transmitter, but RS-485 could manage 32 transmitters and 32 receivers.

With these benefits they were expected to sweep RS-232 aside. But it hasn't happened.

The MAX485 integrated circuit provides both RS-482 and 485 conversions and again only requires a single +5 V supply and operates at up to 250 kB/s.

## UARTs

To convert the parallel data to a serial transmission, we could use a shift register.

The modern alternative is to use a chip called a UART (universal asynchronous receiver/transmitter) or USART (universal synchronous/asynchronous receiver/transmitter).

There are integrated circuits that do the conversion from parallel to serial and back again as a signal is received.

UARTS do a lot more than a shift register. They set the Baud rate for the transmission, they take care of adding parity bits and perform parity checks on incoming signals. They also provide a choice of modulation forms.

A communication system using a UART or USART is shown in Figure 19.3.

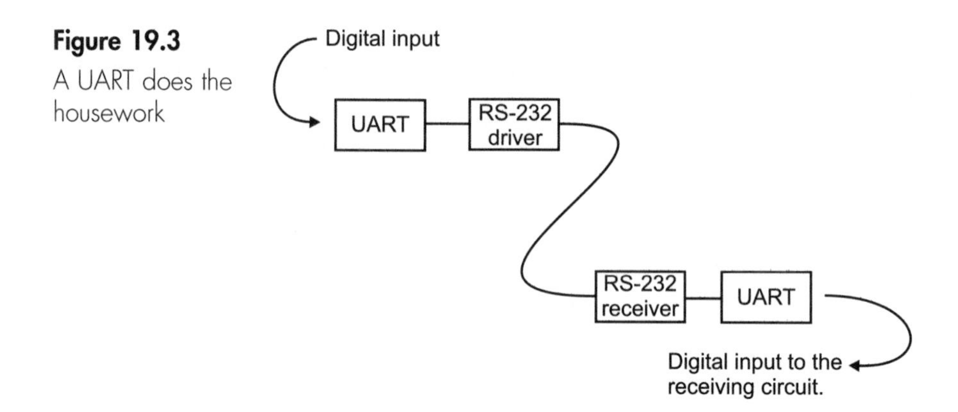


**Figure 19.3**
A UART does the housework

## Encoding

In some systems, the correction of the clock by the start and stop bits only is insufficient and we want to correct the clock more often – once a bit would be nice.

If the data just happens to consist of more positive pulses than negative, the average voltage will have a positive value so we have accidentally inserted a DC voltage into the signal. The value of this DC level is likely to wander slowly between positive and negative with the average value of the incoming data. In some receiving circuits this DC may prove to be a nuisance and cause bits to be missed.

Luckily, both of these problems (the clock synchronism and DC voltages) can easily be solved by making slight modifications to the data before transmission. All we have to do is to 'condition' or 'encode' the digital pulses before they are transmitted, then reverse the process at the receiving end.

### RZ (return to zero)

This is a partial solution.

Every time there is a positive pulse, it is made to drop back to zero during the second half of the clock period. Figure 19.4 shows the original signal, which can be referred to as a 'non-return to zero' or NRZ signal, and its RZ equivalent. The data shown in the figure is just random voltage levels; it is not intended to be RS-232 or anything else.

RZ encoding does not cure the DC level changes although it will reduce the effect, and it will create plenty of clock synchronizing edges when a block of positive voltage occurs but it will not help with a block of 0 V pulses.

### Manchester

This solves everything. Regardless of the data being sent, there is no DC voltage level and every bit of data generates a timing pulse.

The encoding method is to split all bit-times into two. The real data level is sent during the first half of the bit-time, and the opposite level is sent during the second half. So, providing we only read the data during the first half of the bit time, the data is not corrupted at all. The voltage reversal during the second half automatically balances out any DC level, and every bit-time generates an edge to generate a receiver clock pulse. The receiver does not need to generate an accurate clock of its own – it can just derive a clock signal from the incoming data. These waveforms are included in Figure 19.4.

**Figure 19.4**
Two popular encoding methods

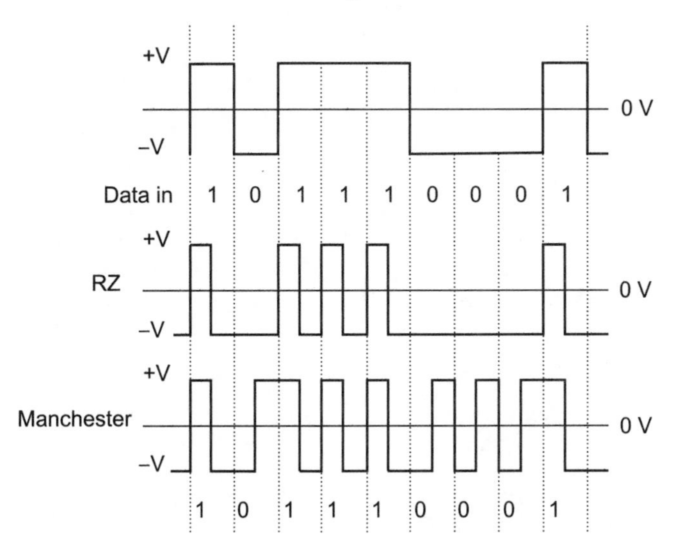

### Multiplexing

This enables us to send more than one digital signal along the same line without them getting muddled up. The same trick works for analogue signals, but that doesn't really concern us just now because we are wearing our digital hats at the moment.

Almost every serious transmission route carries more than a one signal at the same time, and many carry thousands.

### How does it do this?

Easy – at least in principle.

We use a system called TDM – which is an abbreviation for time division multiplexing and not for 'tedium', as some people assume. This just means that we send a small part of the first message then a part of the second message and so on until we have sent some from each message. The circuit that does this is called a multiplexer (MUX). Then we just start again and send some more parts of the messages.

When all these fragments get to the far end of the communication route, another circuit reassembles all the fragments and rebuilds each complete message. This circuit is called a demultiplexer or DEMUX.

### How it works

Figure 19.5 shows a simple mechanical version. In the position shown, a short sample of the logic level in signal A is passed through the switch and sent along the transmission line and, at the far end, is connected back into circuit A.

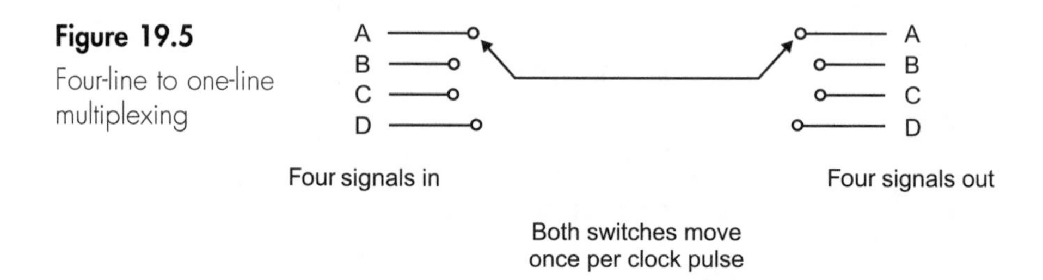


**Figure 19.5** Four-line to one-line multiplexing

Both switches click round to the B position, and much the same happens. The logic level in signal B is sampled, and this information is sent through the transmission line and passed back into circuit B at the far end.

Another click of the switch, and sample C is passed through the switches. This is followed by the same process being applied to signal D.

When this is complete, the switches return to position A and we start all over again.

**How long are the samples and does it matter?**

It is not critical, but there are some guidelines. As we know, the data remains constant for the whole of the bit-time in a transmission and this means that one sample of a pulse is enough. So we don't need to sample each signal more than once per bit-time.

However we must sample all the signals and be ready back at the first one in time to get the data level during the next bit-time.

Using our example in Figure 19.5, if the bit-time is 1 s, the switch will have to move four times in a second in order to be back in time to read the first one again during the next bit-time. The switch would have to move four times in each second, so we would say that the minimum sampling rate is four samples per second.

Each sample would have a maximum width of 0.25 s. Reducing the width of the sampling time, or indeed the gap between the samples, means that the signal on the transmission line would have to change levels in a shorter period of time. This means a higher frequency and a wider bandwidth, which is more expense – so we don't want to go too far down that road. The sampling is shown in Figure 19.6, and has the sample width roughly equal to the gap between samples. This gives the least bandwidth for the system.

**Figure 19.6**
Samples on a TDM system

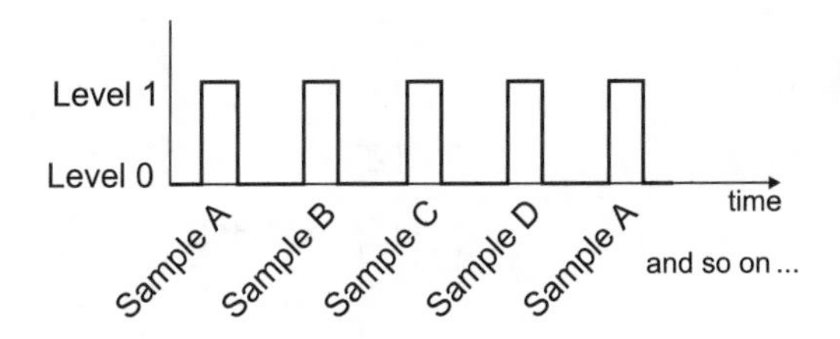


Figure 19.7 shows a situation in which the sample values are A = 0, B = 0, C = 1 and D = 0.

**Figure 19.7**
What data are being transmitted now?

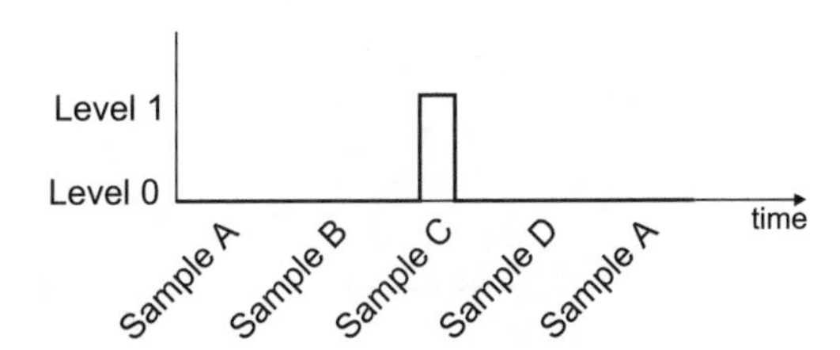

## How do we build it?

Although possible, it is highly unlikely that we will use mechanical switches for all sorts of reasons – speed, switch bounce, electrical noise, life expectancy and cost. A digital solution is altogether better and cheaper.

We need a chip to convert the four incoming lines into a series of samples on a single line. Chips that do this are called multiplexers or data selectors, and there is a wide range available from 2-input to 16-input devices. As an example, we are going to look at the 74XX153. This is a 4-input version.

This chip samples each input in turn, so we must have a mod-4 counter. We can get this by using any binary counter, either in integrated form or by building our own from a couple of JKs.

At the far end of the communication system we need a device to sort the incoming samples back into their four circuits. This chip is called decoder or a demultiplexer. Again there is a range available, from 4–16 output channels. A suitable one to work with the '153 is the 74XX139.

Figure 19.8 shows the data for the 74HC153 and how to connect it to multiplex our four input lines. One thing new is the word 'strobe'. Strobe is just the same as 'enable'. This one is active low.

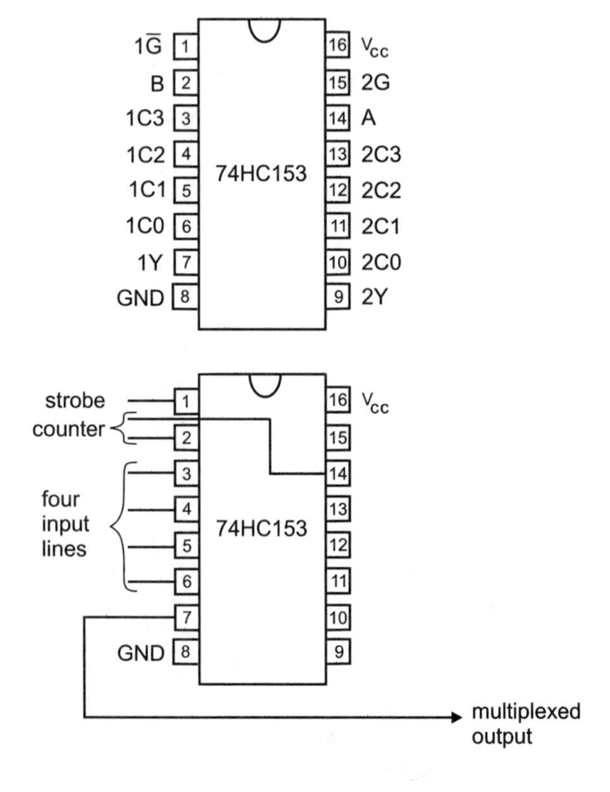


**Figure 19.8**
A multiplexer in use

Figure 19.9 shows the 74XX139 demultiplexer used to recover the four separate sets of data.

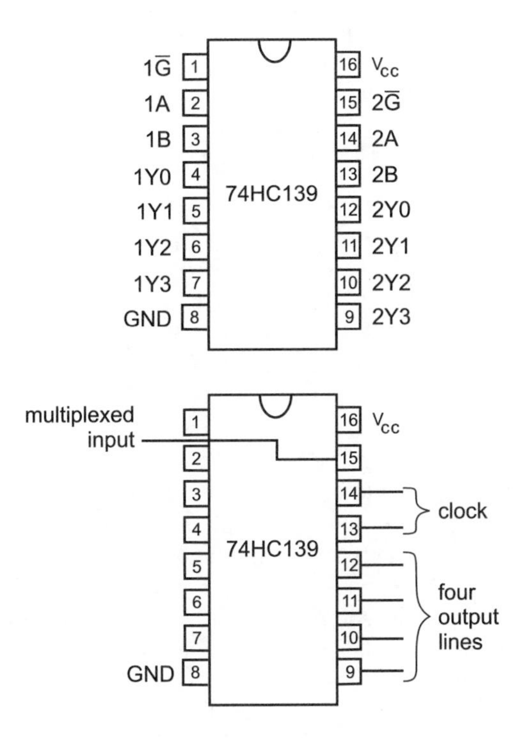


**Figure 19.9**
Using a demultiplexer

The multiplexed input is applied to the strobe input. One small point; the data emerging from the '139 is actually an inverted version of the original input, so it may be necessary to add some NOT gates on the output of the '139.

## Framing

As we discussed earlier, the receiver clock must be synchronized with the transmitter clock. We found that we can do this by encoding with something like the Manchester code to provide plenty of edges.

Even when we have done this, there is another problem. What happens if the receiver misses out a complete sample? All of the succeeding pulses will be synchronized, but would be out of step and sent to the wrong receiving line. In our example, this could mean that line two might receive the data from input line one, line three would get the data destined for line two and so on. In the commercial world this could be dynamite. A company could find that all its research data had safely arrived in a competitor's computer.

To overcome this, we can add a distinctive group that would be recognized by the receiver clock – a bit like adding a bookmark to keep track of how far we have reached in our reading.

A frame then would consist of a framing code followed by the user data (Figure 19.10).

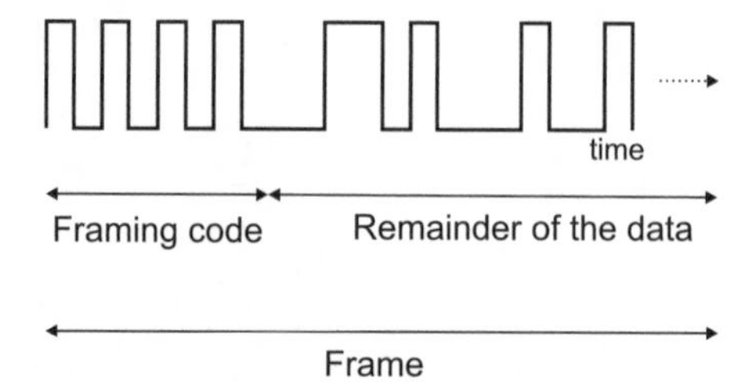


**Figure 19.10**
Synchronizing by framing

**More choices**

The framing code should be a sequence that the receiving circuitry can recognize to enable it to reset the clock. The choice of code becomes more elaborate if the data is very important, because with only 8 bits there is a chance that the chosen code will accidentally appear in the data and cause an unintentional reset. We can make this less likely by increasing the number of bits to 16 or more. Even more sophisticated methods use a sequence of code bits, rather than the same choice each frame.

**How big is a frame?**

Take two extremes:

1 A single framing code at the beginning of the data, and perhaps once a year after that. This is very insecure, but it doesn't add much extra transmission time to the message.
2 The other extreme is to send a framing group just before the data associated with the first transmission line, which in our case will mean an extra group after every four groups of data. This is very safe, but can add 25% to the message.

In between these two extremes, there is one to suit the data transmission being undertaken.

## Modems

Digital signals, because of their fast switching between levels, tend to produce a wide spread of frequencies. The telephone system has a bandwidth of only 300 Hz–3.1 kHz – just about good enough for voice transmission, but not very impressive. To send digital pulses, we employ a modem (MOdulator DEModulator) at each end of the transmission route.

To reduce the bandwidth, the modem converts the logic levels into two different frequency sine waves that account for the noise we hear when the telephone is being used for fax transmissions.

Converting digital signals into different sine wave frequencies is called FSK or frequency shift keying. The waveforms are shown in Figure 19.11.

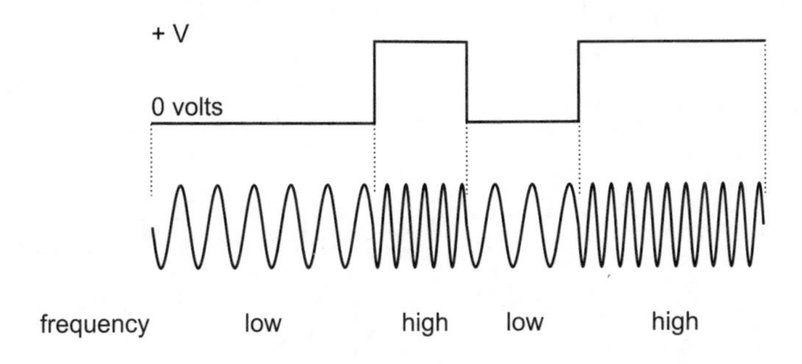


**Figure 19.11** Frequency shift keying as used in modems

## Quiz time 19

In each case, choose the best option.

**1 The data group 10100110110000010010111100 was transmitted using even 1s parity and contains an error. The corrected data group is:**

(a) 00100110110000010010111100
(b) 10100110110000011010111100
(c) 10100110100000010010111100
(d) 10100110110000010010111101

**2 If each channel in a 4-channel TDM system is sampled every 5 ms, the system clock would be running at a frequency of:**

(a) 200 Hz.
(b) 50 Hz.
(c) 10 kHz.
(d) 1 kHz.

**3 Which of these RS-232C output levels would NOT be a valid transmitter output voltage?**

(a) +13 V.
(b) –15 V.
(c) –25 V.
(d) +3 V.

**4 A form of encoding that would be most helpful in clock synchronization is called:**

(a) Manchester.
(b) NRZ.
(c) RS-485.
(d) RZ.

**5 In an RS-232C transmission, the stop bit:**

(a) is sent immediately after the least significant bit.
(b) is not essential if a start bit is used.
(c) is always at a high logic level.
(d) can be longer than any of the other bits.

# 20

# Methods and measurements

## Hooking up a prototype

The first attempt at a new circuit is seldom, if ever, the finished design. It makes sense, then, to build the circuit with a view to making quick and easy modifications without damaging any of the components.

## The prototyping board

This is also called plug board, breadboard and various trade names, including the term 'proto'.

All integrated circuits have pins spaced at intervals of one-tenth of an inch (2.54 mm). For easy building of circuits, prototyping boards consist of rows of interconnected sockets into which nearly all types of components can fit without the need for any soldering. A common layout of these boards is shown in Figure 20.1.

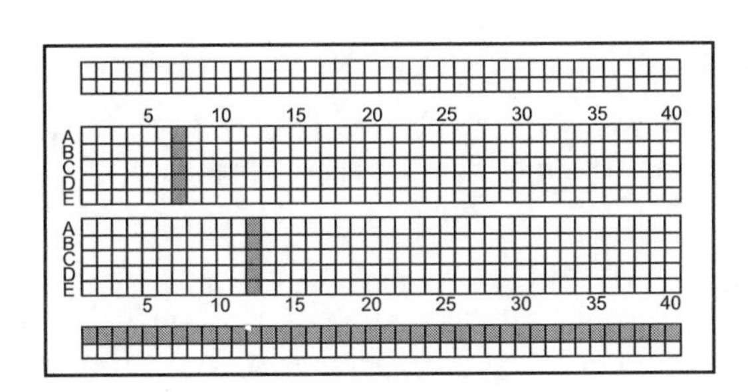


**Figure 20.1**
A prototyping board

Lines of sockets are connected together, some examples of which are shown shaded in the figure. The separate strips running along the outside edges are ideal for running power supplies that are likely to be connected to many parts of the circuits. The ICs, other components and connecting wires are simply plugged in to make the final circuit. A sample circuit is shown in Figure 20.2.

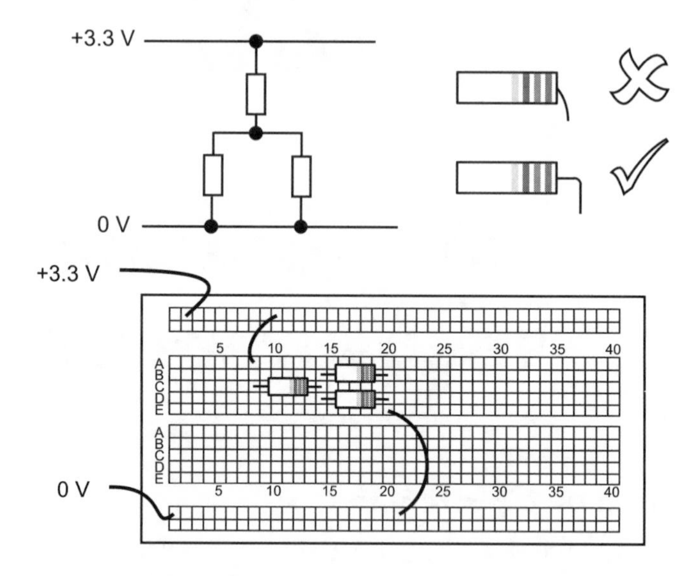


**Figure 20.2**
A sample circuit

**A few tips**

Use single-strand insulated copper connecting wire. A diameter of 0.6 mm is perfect.

Support the component leads on the component side as it is bent to prevent strain and failure. Don't bend closer than one-tenth of an inch from the component.

Use short wires laid flat with right-angled bends, called plate wiring. Avoid high loops of wire that look like a saucer of mustard and cress growing.

Use a method to select different colours of wires to make faultfinding easier – not just to make the circuit look pretty.

## When we are slightly sure of the design

If we want a soldered circuit, we can use stripboard. Vero manufactures much of this, and Veroboard is becoming a generic name for all types of stripboard.

Stripboard is a sheet of insulating material, usually 1.6 mm thick, which has holes punched at one-tenth of an inch centres. On one side, strips of copper provide connections along the board as in Figure 20.3.

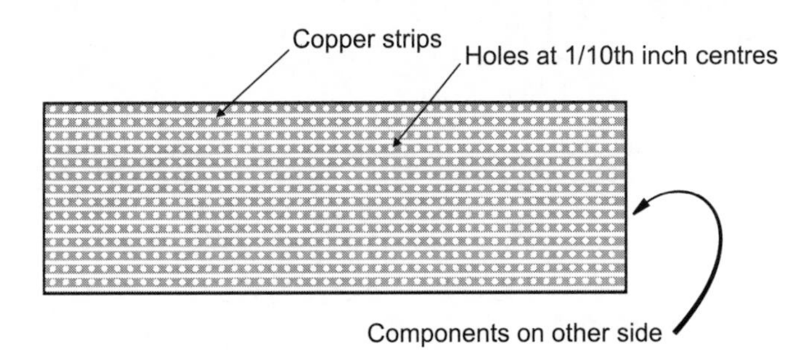


**Figure 20.3**
Stripboard

Components are added on the non-coppered side, and their leads are passed through the holes to be soldered to the copper. Alternatively, pins can be pushed through the holes and soldered in position. The components can be soldered to the pins, making it very easy to try different values with no damage to the copper tracks. Hand tools make the pin insertion easy.

The tracks have to be broken as necessary to allow the circuit to be constructed. To break the track, the best method is to use a Stripboard cutter (also called a strip face cutter). This is basically a twist drill with an attached handle – put it in a hole, press gently and twist clockwise.

## Inserting ICs

We can solder them in just like any other component, but it is not easy to take the IC out again if we change our mind. To remove the IC we either have to melt all the solder at the same time by using one of the special attachments that fit on the soldering iron, or we can use a hand tool called a solder sucker which is like a spring loaded vacuum cleaner. We melt the solder with the iron and, when liquid, we suck it all up, leaving the IC pin clear of solder. Working from pin to pin will finally allow us to lift the IC out.

If the IC is expensive there is another method open to us, and that is soldering a base or socket onto the board so that the IC can be plugged and removed as necessary.

A moments thought is required when selecting the bases to be used. They generally cost at least as much as the IC, and can be up to eight times the price. The very cheap ones are unreliable, have a limited life and are generally a waste of time and money.

The ZIF (zero insertion force) IC base or holder is at the other end of the spectrum. These are excellent to use and a good investment if we are using more expensive chips. We simply lift the lever, which opens the holes, and the pins just drop in. As the lever is lowered the holes decrease in size, gripping the pins. It causes no damage to the IC (Figure 20.4).

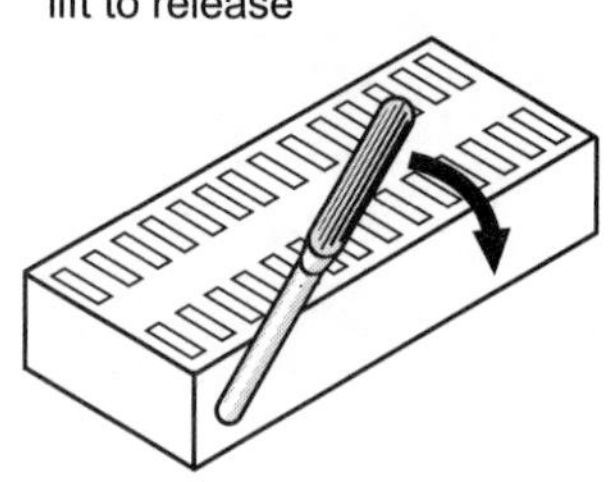


**Figure 20.4**
ZIF – a really nice IC holder

## Printed circuit boards

If we are going to make several circuits of the same design, we may decide to use a printed circuit board (PCB). We can sometimes buy ready-made PCBs for use with some of the more popular chips, or we can make our own.

Basically, a PCB is a sheet of insulator with a layer of copper bonded onto one side. We remove most of the copper to leave behind the connections necessary to build the circuit.

## Construction steps

Draw up the circuit diagram that we wish to build. Build a prototype.

### Design the board layout

Remember that the components will go on the non-copper side of the board. The only other requirement is that it must work. Generally, though, keep ICs in a line with all pin 1s in the same direction, and keep the resistors in line or, if necessary, in groups placed at 90°. Faultfinding becomes more difficult if components are scattered randomly over the board.

Using a sheet of one-tenth inch (2.54 mm) graph paper, lay out the components and mark the component positions in pen. Take a soft pencil and a bountiful supply of erasers and draw in the interconnections necessary to form the circuit, trying not to cross any lines or the copper will be connected causing a short circuit. As the circuit complexity increases, it will become more difficult and probably impossible. We may improve things by repositioning the components.

If connections really must cross, we can use a short length of insulated wire to jump over the offending copper strip. A most impressive way is to hide the jump by using a 0 Ω resistor.

A high-tech alternative is to use a computer and one of the many PCB layout programs, which are really magic to use.

### Transfer the design to the board

The connections on the copper side must be protected from the acid used to remove the remainder of the copper. At this stage take a few moments to add a dot against pin 1 of the ICs, so we don't accidentally plug them in the wrong way round. Instant death for the IC. For the same reason, it is worth marking in the power supply value and polarity.

An easy but not very smart method of transferring the design is to use etch-resist pens to draw it. It certainly works, but the line quality is rather variable.

A better but slower method is to use adhesive tracks or etch-resist transfers. These methods allow us to buy ready-made IC base pads and thus guarantee that the IC will actually fit into position.

The best way is to use a photographic method. To do this we draw the circuit on a transparent polyester sheet, either by applying transfers by hand or by direct printing from a laser printer. This is a nice method; we can go straight from our PC drafting program onto the polyester film.

Using the photographic process, the copper is supplied with an ultraviolet light-sensitive surface coating. The polyester film is placed against this surface coating and exposed to UV light in a light box. **Engage brain at this stage**. Being transparent, the film can easily be placed upside down and the resulting board will be back to front and quite useless. It should be slipped quietly in the waste bin before anyone notices. Alternatively, get someone else to do this bit and have someone to blame. The light softens the exposed areas of the surface coating and the coating under the tracks on the polyester film is unaffected. The softened surface coating is washed off with a developing chemical. At this stage, the now exposed copper is ready for etching.

### Remove the unwanted copper

This is achieved by etching in an acid solution such as ferric chloride. This process dissolves away the exposed copper areas, leaving just the interconnecting tracks.

### Drill some holes

All points where the pins or wires go through the board must be drilled. Small-scale work often uses a small electric drill. If epoxy glass boards are used, spend about three times the price and buy a tungsten-carbide drill bit – well worth the extra. Commercial production is normally by punching the holes through the board. Curiously, this results in a very clean accurate hole with no splintering around the edges.

#### Inserting the components

As the components are inserted, it is usual to bend the leads to about 45° to give some support to prevent the components dropping out as they are soldered. When complete, the excess wire lead should be snipped off and the flux removed. Flux solvents remove the flux easily at this stage, but it gets more difficult as time goes by. There is no strict necessity to remove the flux, but the appearance is much improved and fault-finding is easier.

### Professional boards

As we try to cram more circuit onto a smaller board the PCBs have developed and have more than the single copper layer for the tracks. It is normal now to have multilayer boards with at least four interconnected layers, and sometimes many more. There has also been a move towards surface-mounted devices (SMD). These are small and have no connecting wires. They are mounted by being soldered directly to the surface of the board and very close together, giving a higher packing density.

### Test equipment

We need a basic multimeter to check the input supply voltages, but there are some items that are more specific to digital circuits.

### Logic probe

A logic probe is a simple instrument that has two power connections and a conducting tip which can be touched on points of interest. The general layout is shown in Figure 20.5. There are three LEDs on it. The first two show the logic states 0 or 1, and the third indicates the presence of a high-frequency square-wave or a single very short duration pulse or 'glitch'.

The switch enables the voltage levels to be switched from TTL or CMOS. To ensure the correct interpretation of voltage levels, it is important that the power supply is fed from the circuit that is being tested.

### A logic pulser

A logic pulser acts as a signal generator to inject voltage pulses into a logic circuit. It is very easy to check whether an IC is working by injecting a voltage pulse into an IC and using a logic probe to monitor the changes, if any, at the output.

The input pulses are of very short duration, just a few microseconds, and by providing a current as high as 100 mA it is possible to force the

**Figure 20.5**
A logic probe

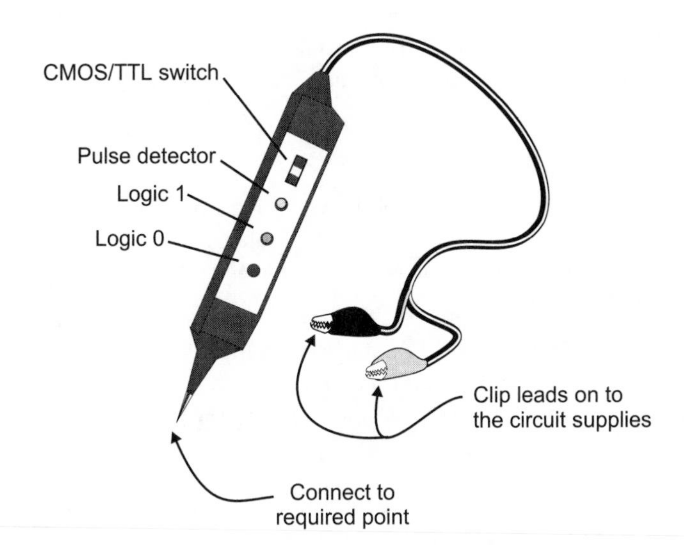


potential to change even if other gates are supposedly holding the voltage at the opposite logic level. Therefore, we do not have to disconnect parts of the circuit before testing. The very short duration ensures that the current is not maintained long enough to damage the circuitry. (Not 100% true, but very nearly!)

They can also provide a continuous square-wave output signal. The frequency is selectable, with a low frequency of about 1 Hz that enables us to watch the circuit operate and hopefully spot the problem. It also has a higher frequency, between 400–1000 Hz, which we can hear with an earpiece as we trace the signal around the circuit.

Physically, the logic pulser looks so similar to a logic probe that we could easily mistake them at first glance.

## A serious piece of test equipment

The previous pieces of test gear would fail when we want to see what data is actually present in a data stream under real operating conditions.

The problem with digital circuits is that they often depend on many bits of data occurring at the right time. We may need to monitor a large number of places, perhaps 30 or more, and then analyse the data slowly and carefully to see what is really happening. This is the function of a piece of test equipment called a 'logic analyser'. Logic analysers are in the same price range as an oscilloscope.

A logic analyser can answer such questions as:

> Are all the correct data inputs present at the instant at which the enable or strobe input is pulsed low?
> Is a dynamic hazard occurring?

The design of a logic analyser is basically a very simple combination of shift registers.

You may remember we looked at shift registers in Chapter 16. The register was loaded with data and, on each clock pulse, the data is moved one place to the left or right as required.

Now imagine a shift-right register that can hold 32 bits of data. If we connect it to a data stream, the first bit of data will enter on the first clock pulse, and after a period of time we will have a copy of the most recent 32 bits of data (Figure 20.6).

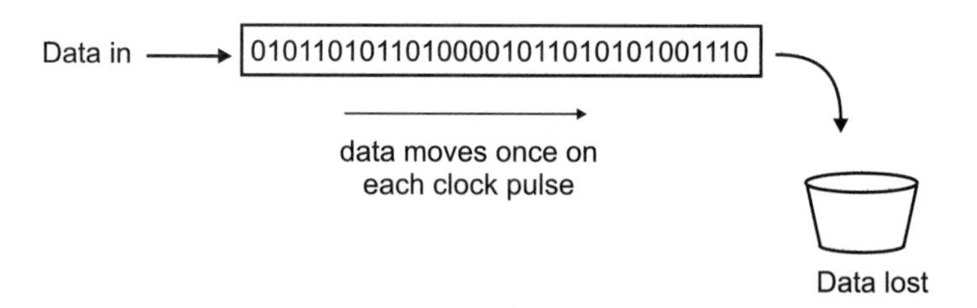


**Figure 20.6** Another look at a shift register

Now, if we had four such registers, we could collect data from any four parts of a circuit at the same time.

Imagine we were trying to locate a fault in a counter. We may want to know how the output from each JK compares with the clock input. We would use one register to monitor the logic levels on the clock, and three others to check the outputs.

The central part of the register is called a 'window', and allows us to read the data stored at that point (Figure 20.7). A simple arrangement

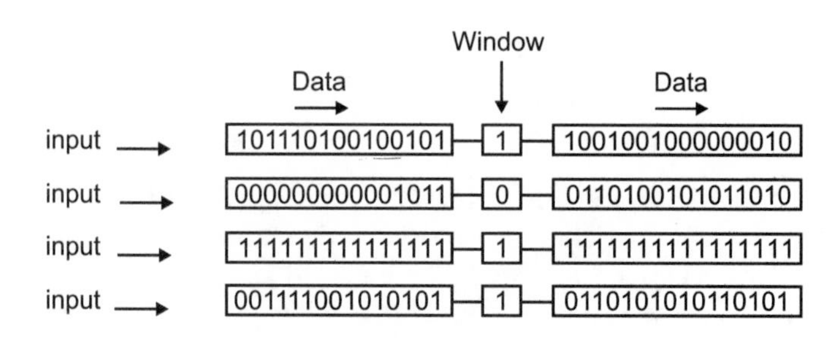


**Figure 20.7** The window value is 1011

like this would be referred to as a 4 × 32 (4 by 32) logic analyser. In logic analyser specifications, the number of registers are spoken of as the number of channels.

In real life, we would never find such a simple arrangement of registers. Logic analysers contain at least 32 channels, each containing a 2048-bit shift register. One like this is referred to as a 32 × 2k logic analyser, and with it we can monitor any 32 different points in a digital system. Which points we choose are entirely up to us.

### More about the window

In the window, the logic analyser will 'see' a bit of data from each of the channels. We can also load a combination that we are searching for. For convenience we can enter the values in hex or any other base, and as the clock pulses arrive from the circuit the data move across and are continuously compared with the combination we have entered. When a match is found, the clock is switched off and the data 'captured'.

We can now move backwards and forwards along the registers and read off the sequence of data bits 'frozen' in time, so we can analyse it at our leisure. The data can also be displayed as waveforms on an oscilloscope.

The benefit of positioning the window in the centre of the shift register is that it allows us to observe the program action before as well as after the chosen moment.

## Quiz time 20

In each case, choose the best option.

**1 Integrated circuits have pins spaced:**

(a) 1 mm apart.
(b) one-tenth of an inch apart.
(c) 1 m apart.
(d) 2.54 in apart.

**2 The letters ZIF stand for:**

(a) zero insertion force.
(b) zoom insertion fixing.
(c) zero integrated force.
(d) zero insertion fixing.

**3 Most single-layer PCBs have a thickness of:**

(a) one-tenth of an inch.
(b) 0.6 mm.
(c) 2.54 mm.
(d) 1.6 mm

**4 An advantage of using SMDs is:**

(a) that the larger components make handling easier.
(b) their higher power dissipation.
(c) that a higher packing density is achieved.
(d) that the copper can be etched without ultraviolet light.

**5 A device used to inject a voltage into a digital circuit is a:**

(a) logic pulser.
(b) flux meter.
(c) logic analyser.
(d) logic probe.

# 21

# Avoiding some problems – and finding the others

I built my first radio using the techniques illustrated in Figure 21.1.

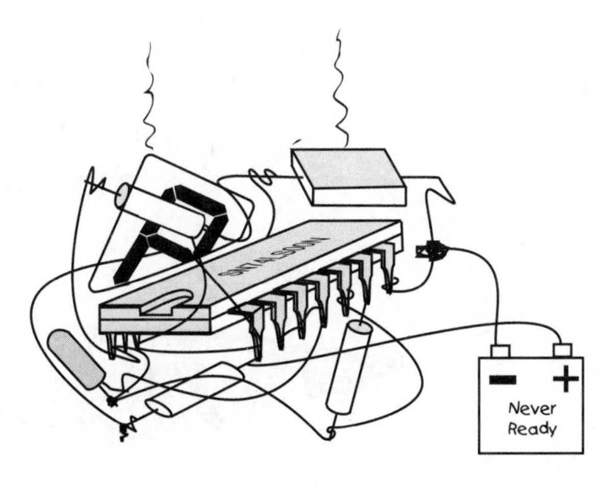


**Figure 21.1**
Look Mum, I built it myself!

I sat up through the night, determined to finish it. I held all the circuit in my head. Finally I connected the last wire and switched on. It didn't work. It took me a fortnight to remove all the faults I had introduced. Then it worked.

It was a waste of time and effort but it taught me a lesson or two.

## Ways to avoid problems.

1 Keep notes during building or modifying a circuit, otherwise the telephone will ring or the fire alarm will sound and 10 minutes later we are staring at the circuit with no idea of how far we have got.
2 Add labels and colour code the wiring. Following a circuit with inconsistent colour coding is a nightmare.
3 Always make sure the power supplies are applied before the input data is connected. It is often fatal for an IC to have input voltages applied without the supplies.
4 Never remove or insert any component or plug or IC with the supplies on. Inevitably some connections are made or broken before others, so voltage changes occur in a random order.
5 Don't cook components with the soldering iron. They don't like it.
6 Don't break the ICs. Take them out of the bases carefully. This is sometimes easier said than done. The recommended way is to invest in an IC remover tool that grips the chip along the whole of its length, allowing it to be eased out of the socket without breaking. The very popular but dangerous way is gently to prise up alternate ends with a small screwdriver until it is lifted clear of the base. As the size of the IC (and usually its price) increases, the chance of cracking it in half becomes more serious. Safer than a screwdriver is the cap of a ballpoint pen, as in Figure 21.2. Just slide it under the chip and wedge alternate ends up. The softness of the plastic reduces, but does not eliminate, the chance of damage. There are also bases with a lever, very similar to ZIF bases, which actually lift the IC out when the lever is operated.

**Figure 21.2**
Safer than a screwdriver

7 Remember electrostatic discharge (ESD). The input resistance to CMOS gates is extremely high, which means that static electricity applied to the pins of the IC cannot leak away easily and quickly builds up a voltage sufficient to damage the chip. The earlier gates like the 74 and 74LS families have much less trouble with static.

Nowadays the only safe way is to assume that all ICs are static-sensitive.

## Electrostatic discharge

The increase in speed and complexity over the years has resulted in using smaller internal geometry, which in turn means thinner layers of insulation and more susceptibility to ESD damage.

We cannot feel a static discharge unless it exceeds 3000 V, but 200 V of static electricity is enough to damage chips. This is despite the fact that most chips have antistatic protection circuits at their inputs. We can certainly improve the antistatic precautions and indeed we can cure the problem completely, but only at a cost of a large reduction in operating speed.

Static electricity can degrade the performance without actually killing it totally. In fact, it can cause any number of odd effects.

What can we do about it? This is not an easy question to answer – we could stop it altogether, but do we really want to? A better question is 'what steps are worth taking, given all the circumstances'? We have a close parallel on the roads – we can totally eliminate serious road accidents but have decided not to. It would be easy. Fit hydraulic rams to the front and rear of all vehicles to cushion any impact, fit Doppler radar to apply the brakes automatically when any obstruction or road problem is sensed ahead of the vehicle, and reduce the maximum speed to 5 mph. Problem solved.

## What steps shall we take to protect a few integrated circuits?

Circumstances alter cases. Consider, for instance, assembling a circuit using ICs costing over a hundred pounds each as part of a 'missile launch inhibit' circuit. Under these circumstances, it is worth taking the most stringent precautions. On the other hand, if we are using a single JK chip to make a counter just for fun, the simplest precautions will be enough.

So, choose from the following.

### Simple precautions at no cost

1 Store the chips in conductive foam, aluminium foil, antistatic plastic bags or antistatic bubble wrap.
2 Check the furniture. Avoid plastic chairs or tables and nylon carpets. A bare wooden table is very safe, but beware of a varnished table. It may appear to be wood, but if it is coated with polyurethane varnish then, in the opinion of the static electricity, it is a plastic table.

3 Check clothing. A nylon lab coat is bad news – particularly if you are in a low-humidity area due to central heating or air conditioning. Cotton is very much better, as it absorbs moisture more easily and becomes a poor insulator.
4 Check the floor. Avoid plastic tiles unless they are designated antistatic.

### More serious precaution

1 Use a wrist strap. This is a conductive band that fits round the wrist and is earthed by a cord clipped to a convenient earth point. They have a resistance of about 1 MΩ. This conducts enough to dissipate static, but not enough to electrocute us if we accidentally touch something nasty. A very worthwhile first step to safety, they are quite inexpensive and are sometimes provided free of charge when buying the more expensive chips.
2 Conductive matting is available to put over the surface of the workbench and on the floor under the chair. This is also earthed by a conductive lead, as on the wrist strap.
3 A conductive heel strap provides a connection between the user's leg and the bottom of the shoe to allow easy movement in areas like storerooms or production lines.
4 All chip storage facilities are available in an antistatic form.

## Fault-finding

Good fault-finding does not involve a great deal of activity; we don't have to keep up an average of a check a minute to show that we are apparently 'getting on with it'. In fact fault-finding is not like that; it involves a lot of time sitting and thinking. (Not just sitting).

Whenever we make a test, we should always write down the test and the result. This is vital, as it saves us repeating tests that we have already made. It also allows us to settle down with a cup of coffee and sift through the clues we have gathered and plan our next move.

We must follow a method, and here are the two popular ones.

### The half-split method

This is probably the fastest way of homing in on a fault. The first test is made at a convenient point in the centre of the circuit, as shown in Figure 21.3.

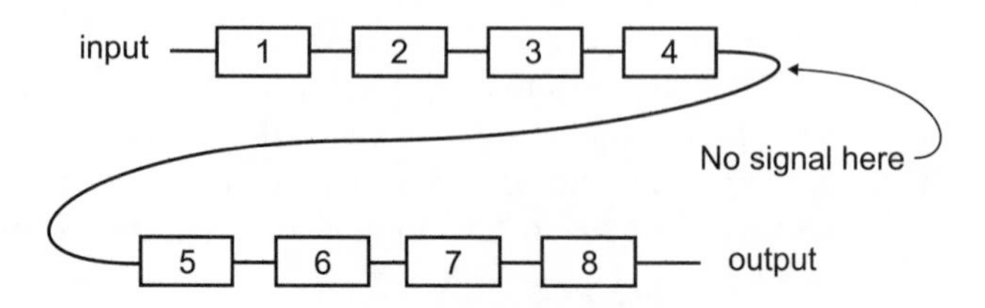


**Figure 21.3** The half-split method

If the expected signal is not present the fault must be somewhere in blocks one to four, so the shaded blocks five to eight have been eliminated. Our next step is to check halfway through the remaining blocks, as in Figure 21.4.

**Figure 21.4**
Test halfway through the remaining circuit

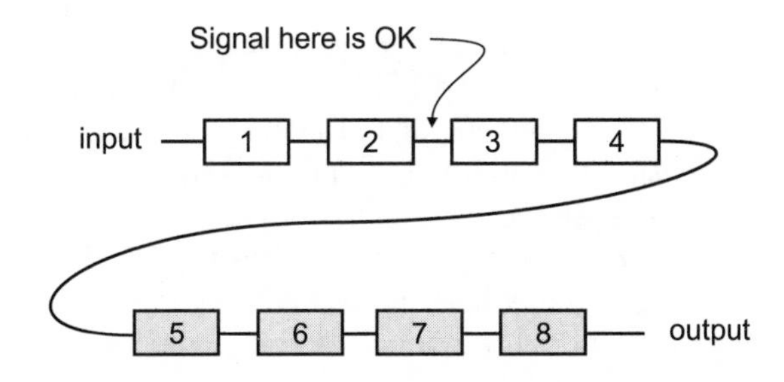


If we assume that a signal is present at this point we know that the circuit elements one and two are OK, so again we make a check between the two remaining parts of the circuit as in Figure 21.5.

**Figure 21.5**
One last test

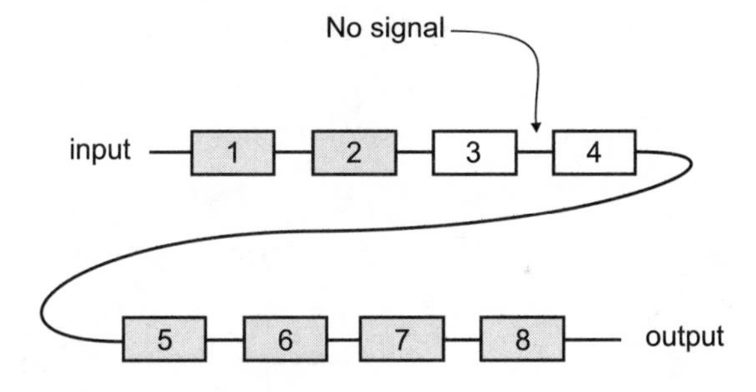


If no signal is found, then the fault must be in block three (Figure 21.6).

**Figure 21.6**
Got it!

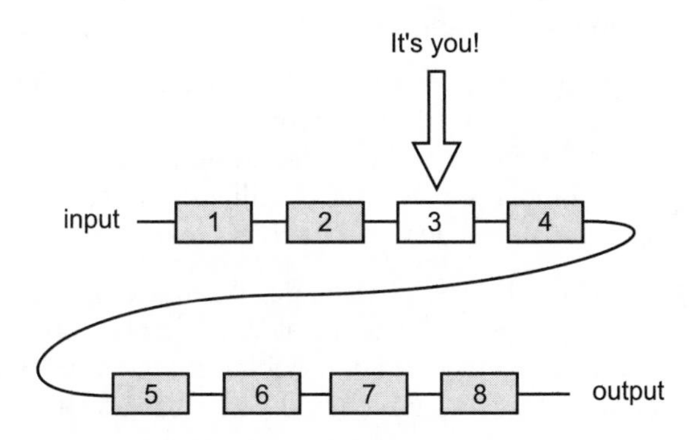

## Plod-through method

This means starting from the beginning and checking the output of each stage, following the signal until, at the faulty stage, it disappears. The alternative is to start from the last stage and check back through the circuit until the expected signal is found.

People who like the plod-through approach often have a firm opinion about which one is best. Those that prefer to start at the beginning have the benefit of following a known signal, so test equipment is easier to set up and any loss of signal is immediately apparent. However, working back from the end means that we are moving from a 'no signal' situation to 'some signal'. The advocates of this method would argue that it is fast and easy to spot a signal, and we can always worry about the details of the signal when we have got one.

At the end of the day it doesn't matter which way we prefer. Any method is better than no method.

## Checking for a waveform

Even when we are following a recognized method, we still have to keep alert. In the last example, we checked between blocks three and four and found no signal. We had 'found' the fault to be block three.

Before rushing out and buying a new chip, let's look more closely. The actual circuit is shown in Figure 21.7.

**Figure 21.7**
Which gate is dud?

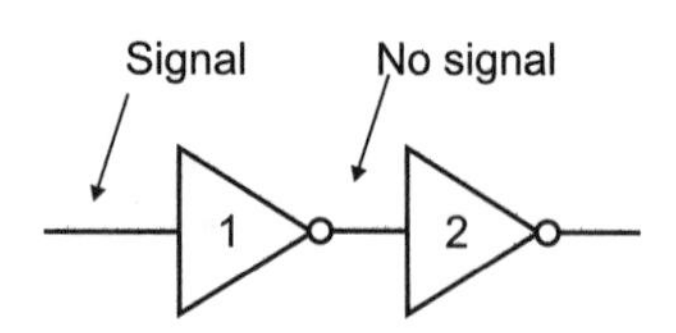


We check the power supplies to both chips and all supplies are OK. We break the link between the two chips and the signal at the output of chip 1 mysteriously reappears. There had been a low resistance path to the 0 V supply inside the second chip.

Breaking the link could be done by removing the second IC from the board or by disconnecting the wire or PCB track. This should be achieved by using a hand tool called a track cutter, but many technicians resort to a drill or scratching away with a knife or a screwdriver until the track is broken.

When the repair is made, the track should be cleaned and a short length of copper wire soldered across to bridge the gap. Try to avoid the quick solution of blobbing solder across the gap. This often results in a dry joint, which will come back to haunt us later.

## Checking an IC

Checking by replacement is an easy way of knowing for sure whether it is the IC causing the problem. But there are a couple of provisos here. We must know that the IC used for the check is not faulty. New chips are not fully checked after manufacture. Generally only a sample of each batch is tested, so some duff ones always get through. Some chips like RAMs cannot be conclusively checked. They are tested by loading 1s into each memory cell, then by loading 0s in each. Finally, 0s and 1s are loaded into alternate memory locations; other data combinations are not tested. We do not have a computer in the world that is fast enough to test all combinations of data in a simple RAM chip.

The other problem is that we must run the risk that if there is a circuit fault that caused the failure, it may kill off our replacement as well.

It is a good idea to check the pin voltages first, just looking for something really silly. The likely pins that are worth checking are the ones carrying a DC voltage, like the power supplies. This may show a fault in the PCB track layout.

## Reliability of components

In order, worst first, the least reliable components are connections and cables, switches and variable resistors.

## Probable causes of IC faults

I analysed the causes of a large number of faults occurring in digital circuits that employed integrated circuits.

There were seven main causes that account for almost every failure of digital circuits, and these are listed below.

### No power supplies connected to the chip

Magic they may be, but they won't run without supplies. They are sometimes forgotten at the prototype stage, as most circuit diagrams omit the power supplies for clarity on the page.

### Miscounted pins

Remember the pins are counted from above the IC around it in an anticlockwise direction (Figure 21.8).

**Figure 21.8**
Count the pins

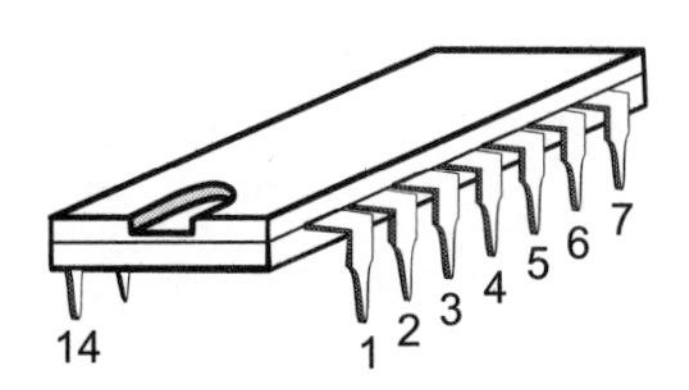

### Inputs left floating

Floating inputs tend to float up to logic 1 levels and sometimes the circuits still work but are prone to erratic action due to noise. These circuits are often reported as 'it was working yesterday/before lunch/ just now but it's decided to go wrong'.

When using a voltmeter they show 0 V between the floating input and ground. They also show zero between the positive supply voltage and the suspect pin, showing that something is clearly amiss (Figure 21.9).

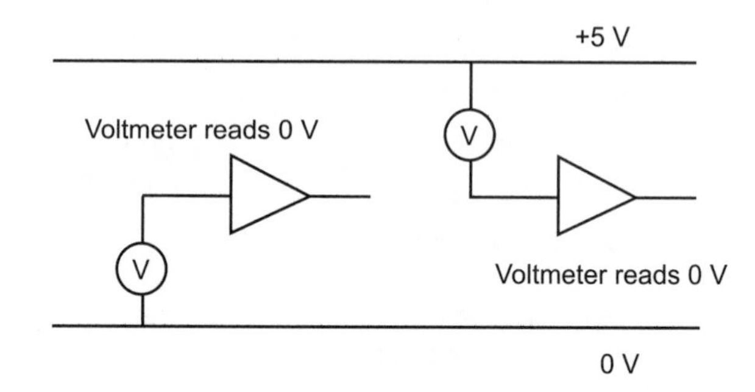


**Figure 21.9**
Strange but true

A logic probe would not be fooled by a floating 'zero'; it will not show any logic level if it is floating.

### Insufficient decoupling

This is another cause of erratic running, and was introduced in Chapter 11.

### Bent pins

If the pin is bent when being inserted into a base it is very difficult to see, and if the pin happens to be an output pin the test instruments will show the output to be OK. See Figure 21.10.

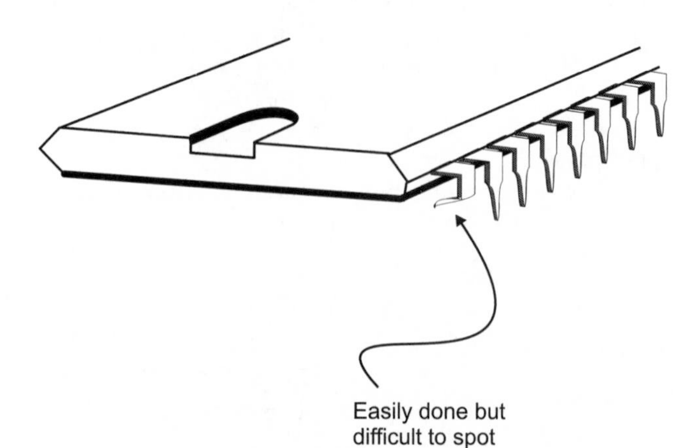


**Figure 21.10**
Bent pins can cause puzzles

### IC inserted the wrong way round

This normally reverses the polarity of the supplies. The chip normally gets very hot and does not survive the experience. To help us to avoid this, a dot is printed on the board next to pin 1.

### Voltages applied to the output

If a chip is trying to make its output go to a logic 0, say, and we apply +5 V to it, there will be a bit of a disagreement and the result may well be a dead IC.

## Feedback loops

In circuits like mod-n counters, where the output controls the operation of the circuit, a failure in any part of the circuit will cause the whole thing to collapse. It may be an idea to disconnect the feedback section and look to see if the circuit performs in the expected minimal way without it. With a counter, for instance, it may just count continuously without resetting at the required value. It may be worth taking all the enable or strobe inputs to their active values and seeing if the outputs then respond correctly. It may also be possible to slow the input clock pulses by replacing the input clock with a debounced switch. We can then check the operation of the circuit on every clock pulse.

Finally, late at night, if all else fails, just switch it off. Go outside and stare up at the stars. They will tell us the truth – in the global view, our problems really don't matter.

## Quiz time 21

In each case, choose the best option.

**1 The half-split system of fault-finding:**

(a) is probably the fastest way of finding a fault.
(b) can only be used in circuits which include feedback.
(c) involves breaking each integrated circuit in half.
(d) can be started at either end of the circuit.

**2 The best way to locate a floating input is by using:**

(a) a voltmeter.
(b) an oscilloscope to check the output of the chip.
(c) an LED.
(d) a logic probe.

**3 An integrated circuit is most likely to be damaged by:**

(a) leaving an input pin floating.
(b) reversing the power supplies.
(c) disconnecting an output pin.
(d) disconnecting the input voltages before disconnecting the power supplies.

**4 The least reliable parts of an electronic circuit are:**

(a) 1 MΩ resistors.
(b) connections and cables.
(c) decoupling capacitors
(d) integrated circuits.

**5 A gate with a correct input but no output:**

(a) must be replaced.
(b) is not faulty, it is in a circuit which includes feedback.
(c) may not be faulty.
(d) has overheated but will recover if left to rest.

# Quiz time answers

### Quiz time 1

1 (c)
2 (a)
3 (c)
4 (a)
5 (d)

### Quiz time 2

1 (c)
2 (b)
3 (d)
4 (d)
5 (b)

### Quiz time 3

1 (d)
2 (b)
3 (a
4 (a)
5 (d)

### Quiz time 4

1 (c)
2 (a)
3 (d)
4 (b)
5 (c)

### Quiz time 5

1 (d)
2 (a)
3 (c)
4 (a)
5 (c)

### Quiz time 6

1 (d)
2 (b)
3 (b)
4 (c)
5 (d)

## Quiz time 7

1 (d)
2 (a)
3 (c)
4 (a)
5 (b)

## Quiz time 8

1 (a)
2 (b)
3 (c)
4 (b)
5 (a)

## Quiz time 9

1 (b)
2 (c)
3 (d)
4 (a)
5 (c)

## Quiz time 10

1 (d)
2 (a)
3 (b)
4 (a)
5 (c)

## Quiz time 11

1 (c)
2 (c)
3 (a)
4 (a)
5 (b)

## Quiz time 12

1 (a)
2 (c)
3 (a)
4 (d)
5 (a)

## Quiz time 13

1 (d)
2 (b)
3 (a)
4 (c)
5 (c)

## Quiz time 14

1 (c)
2 (c)
3 (b)
4 (d)
5 (a)

## Quiz time 15

1 (a)
2 (d)
3 (c)
4 (d)
5 (b)

## Quiz time 16

1 (a)
2 (c)
3 (d)
4 (a)
5 (b)

## Quiz time 17

1 (c)
2 (a)
3 (d)
4 (b)
5 (b)

## Quiz time 18

1 (d)
2 (b)
3 (a)
4 (c)
5 (c)

## Quiz time 19

1 (d)
2 (b)
3 (c)
4 (a)
5 (d)

## Quiz time 20

1 (b)
2 (a)
3 (d)
4 (c)
5 (a)

## Quiz time 21

1 (a)
2 (d)
3 (b)
4 (b)
5 (c)

# Further Reading

Horowitz, P. and Hill, W. (1989) *The Art of Electronics*. Cambridge University Press, Cambridge.
Texas Instruments (1997) *Selection Guide and Databook* (CD-ROM).
Schweber, W. L. (1988) *Data Communications*. McGraw-Hill.

# Index